T0135009

Studies in Computational Intelligence

Volume 513

Series Editor

Janusz Kacprzyk, Warsaw, Poland

For further volumes:
http://www.springer.com/series/7092

Amelia Bădică · Bogdan Trawiński
Ngoc Thanh Nguyen
Editors

Recent Developments in Computational Collective Intelligence

Editors
Amelia Bădică
Faculty of Economics
and Business Administration
University of Craiova
Craiova
Romania

Ngoc Thanh Nguyen
Institute of Informatics
Wroclaw University of Technology
Wroclaw
Poland

Bogdan Trawiński
Institute of Informatics
Wroclaw University of Technology
Wroclaw
Poland

ISSN 1860-949X ISSN 1860-9503 (electronic)
ISBN 978-3-319-03331-0 ISBN 978-3-319-01787-7 (eBook)
DOI 10.1007/978-3-319-01787-7
Springer Cham Heidelberg New York Dordrecht London

Printed on acid-free paper

Springer is part of Springer Science+Business Media (www.springer.com)

Preface

Collective Intelligence represents the intelligence shared by a group of intelligent entities (natural or artificial) resulted from their mass interaction through various activities including collaboration, competition, opinion expressing, value exchange, message exchange, a.o. The term occurs in sociology, populations' biology, economy and political sciences. This interaction can involve for example consensus reaching, decision making, social choice or other means for quantifying mass activities. The rapid growth of the interconnectivity of a huge number of intelligent natural and artificial entities powered up by the most recent Internet and Web technologies (the Internet of Things, the Web of Services, the Web of Data, the Semantic Web, and the Social Web) will determine the ad-hoc formation of heterogeneous intelligent complex systems that massively combine natural and artificial intelligence resulting in new forms of Computational Collective Intelligence.

Computational Collective Intelligence is a rapidly growing field that is most often understood as an AI sub-field dealing with soft computing methods which enable making group decisions or processing knowledge among autonomous units acting in distributed environments. Web-based Systems, Social Networks and Multi-agent Systems very often need these tools for working out consistent knowledge states, resolving conflicts and making decisions.

The papers included in this volume cover a selection of topics of the rapidly advancing domain of Collective Computational Intelligence: formal computational models of markets and social systems, emergent behaviors, autonomous agents, planning, agent-oriented programming, character recognition, intelligent transport, neural network applications, optimization, grey theory, natural language processing, group decision support, and rule systems.

The book consists of 19 extended and revised chapters based on original works presented during a poster session organized within the 5th International Conference on Computational Collective Intelligence that was held between 11 and 13 of September 2013 in Craiova, Romania. The book is divided into three parts. The first part is titled "Agents and Multi-Agent Systems" and consists of 8 chapters that concentrate on many problems related to agent and multi-agent systems, including: formal models, agent autonomy, emergent properties, agent programming, agent-based simulation and planning. The second part of the book is titled "Intelligent Computational Methods" and consists of 6 chapters. The authors present applications of various intelligent computational methods like neural networks, mathematical optimization and multistage decision processes in areas like cooperation, character recognition, wireless networks, transport, and metal structures. The third part of the book is titled "Language and Knowledge Processing Systems", and consists of 5 papers devoted to processing methods for knowledge and language information in various applications, including: language identification, corpus comparison, opinion classification, group decision making, and rule bases.

The editors hope that this book can be useful for graduate and PhD students in computer science as well as for mature academics, researchers and practitioners interested in the methods and applications of collective computational intelligence in order to create new intelligent systems.

We wish to express our great attitude to Prof. Janusz Kacprzyk, the editor of this series, and to Dr. Thomas Ditzinger from Springer for their interest and support for our project.

The last but not least we wish to express our great attitude to all authors who contributed to the content of this volume.

Amelia Bădică
Bogdan Trawiński
Ngoc Thanh Nguyen

Contents

Part I
Agents and Multi-agent Systems

Formal and Computational Model for A. Smith's Invisible Hand Paradigm

Tadeusz (Tad) Szuba, Stanisław Szydło, and Paweł Skrzyński

AGH University,
Dept. of Applied Computer Science, Faculty of Management
{szuba,skrzynia}@agh.edu.pl, sszydlo@zarz.agh.edu.pl

Abstract. This paper presents a formal theory of what A. Smith's Invisible Hand is. We are convinced that the Free Market as a system of intelligent, interacting, selfish agents creating a powerful and universal computational mechanism. The model of computations for this mechanism is based on three concepts: molecular computations, implied computations and value labeling logic expressions representing business objects. This runs threads of computations in the background of agents' consciousness. The threads are chaotic, parallel, distributed computations of inference nature. The information stored and processed by threads is labeled with "value", assigned and changed in subjective way by agents - to control the inference process. The results of inferences, affect the Free Market, and are perceived as A. Smith's Invisible Hand, or Free Market Collective Intelligence. Research suggests that this phenomenon is wider than A. Smith expected and more aspects of the Free Market can probably be explained. This paper is the first presentation of our research. The simulation model along with results obtained will be demonstrated during the conference.

Keywords: A. Smith's Invisible Hand, Collective Intelligence, Free Market self-regulation, molecular model of computations, reasoning in logic.

1 Introduction

Since A. Smith coined the term "invisible hand" e.g. [8], we became aware that the Free Market (FM) is able to self-control. This self-control is so advanced and unpredictable, that man is unable to take full advantage of this awareness. This self-control seems to be more "intelligent" globally than any ruling economist or team of economists. All new theories trying to deal with this phenomena, e.g. "fractal market analysis" [5] usually after a decade or so, lead again to a conclusion that we do not fully understand the Free Market (FM). Economic cycles and crises are sufficient proof of this. The theory presented here assumes, that A. Smith's Invisible Hand (ASIH) is "a product" of certain Intelligence, which is different from human intelligence. It is "Free Market Collective Intelligence" (FMCI). In this paper, we will present a formal model of FMCI, and show the nature of computational processes behind it. The physical layer and key components of corresponding computational model (much like neurons for humans, transistors for computers) will be explained. In our life we frequently face situations where we do something, and later we conclude

A. Bădică et al. (eds.), *Recent Developments in Computational Collective Intelligence*,
Studies in Computational Intelligence 513,
DOI: 10.1007/978-3-319-01787-7_1, © Springer International Publishing Switzerland 2014

that something other, unplanned, has been generated. A good example is a situation where a piece of hardware is produced for certain application and later, an unexpected application for its use has been found. Let's name such a case "implied tool"; because when created, the second application is implied, "outside of our consciousness". This case can be recursive. This can also happen in case of computations when e.g. Java sub-classes can be used for different applications. Such classes can be part of much bigger, implied system of computations. Section 2 will demonstrate implied computations even in a simple Turing Machine. In an ASIH model of computations, inferences are locally consistent, however main control is based on *value* arbitrarily assigned to elements of inference, derived by an agent in subjective way from a Free Market situation. Such a *value* can change during the inference process. It can be propagated along the inference net. Our theory states that personally[1] intelligent and selfish Free Market agents (persons or companies) thinking and performing Free Market activities, create unconsciously create a "hardware platform" where distributed, parallel and chaotic computations of inference nature take place, which we perceive as FMCI. From a computational point of view, a single abstract agent active in a Free Market should at least:

- Permanently process all available market information because in general, the agent cannot predict which information will be helpful for future business;
- Be able to build abstractions over the real world in such way, that logic expressions additionally labeled with subjective *value*, are assigned to real objects and actions. Theoretical considerations on this concept can be found in [3];
- Be able to define the present *value* of market item, and to predict possible future *value* of items after assumed transformation. Thus, *value*[2] is not a price: *value* is of logical nature used to label logic expressions, but price (measured with the help of money) is of economic nature. The transaction price for specific item can be derived from entry *values* assigned by both: buyers and sellers to the item during the process of bargaining;
- Be able to build and evaluate possible sequences (nets) of actions, such that the final sum of costs of single steps in the sequence is essentially smaller than final business profit;
- Finally, an agent realizing a business plan must be able to locally change the FM.

The above list of required abilities and actions implies huge, redundant mental computational power of agents, which can easily host additional, unconscious computations (inferences) constituting FMCI. Single agents thinking about own business, do not realize, that they also participate in threads of inferences related to FMCI.

The rest of the paper is structured as follows; In Section 2, the state of art is presented. Section 3 explains how to understand self-regulation of FM. Section 4 and 5 introduces the concept of "implied computations". Another example is given. Section 6 describes a simulation model used to justify our theory. The last section contains conclusions.

---

[1] Human intelligence must reach level of 3-years children, to allow it to understand the nature of money in our life. At this moment children reaches mental level of chimpanzee [11].

[2] The difference between value vs. price is discussed in the book: K. Marx: Das Kapital.

2 State of the Art

Despite many attempts [4], [6], [7], [13] at designing market simulation models, not all can be considered convergent with our research. Some of the studies are focused on developing a new approach to the problem of allocating resources based on abstractions, taken from economic theory (such as the laws of supply and demand) and computer science (multi-agent systems) [9], [13]. Other research presented the ASIH by reference to the equilibrium price problem in the evolutionary market theory [6]. The design of economic simulators was also attempted for the purposes of computer games[3]. This section discusses previous attempts to simulate the market, particularly approaches based on multi-agent systems, which the authors consider close to this work. These approaches fall within the direction of research referred to as the Agent-Based Computational Economy (ACE)[4]. However, the approach presented by the authors is more „object" oriented in the sense of making use of real market elements. Combining these with Collective Intelligence, makes deriving abstract computational processes from the real market possible. An interesting tool developed as part of ACE is REPAST (*REcursive Porous Agent Simulation Toolkit*): an agent framework created for the purposes of sociology and available free-of-charge. REPAST makes it possible to systematize and research complex social behavior by offering a functionality for designing controlled and repeatable computational experiments. It is notable that the intended purpose of this framework was much more general and not restricted to economics. Another interesting project initiated as part of ACE is the JAMEL project, which is a distributed macro-economic simulator implemented in Java. The model used in JAMEL boils down to a multi-agent system where money is the endogenous variable. Market simulators designed with the involvement of scientific and commercial institutions support scientists and decision makers to the following extent:

- Researching the market reaction to various, frequently quite improbable events, such as analysis of consequences of a hypothetical explosion of prices of water, food, oil or other goods;
- Predicting reactions to individual business decisions: even small steps taken by boards of large corporate groups can significantly and often irreversibly impact the standing of the corporation and its affiliated companies, so simulating the market response to individual decisions allows some potentially dangerous errors to be avoided;
- Empirically checking the accuracy of existing economic models describing the market: the simulation can be used to confirm, correct or reject hypotheses.

In addition, division lines based on the nature of data contained in the system and the method of managing market agents can be drawn. Due to the nature of data contained in the system, the following are distinguished:

---

[3] It is worth noting such games with cult following today as Simcity or Civilization.

[4] A very good website of Professor Leigh Tesfatsion of the Iowa University which can form a starting point: `http://www.econ.iastate.edu/tesfatsi/ace.htm`

- Fictitious simulators operating on abstract or untrue data;
- Real simulators operating, for instance, on the real volatility of share prices or exchange rates, employed particularly often for running equity investment games without using real money.

Depending on the method of managing market agents, the following are distinguished:

- Automatic management – decisions of the majority of agents are made automatically following defined reasoning rules;
- Manual management – the majority of agents take decisions in accordance with instructions from equity investors.

The simulators depend greatly on the model (theory) forming the foundation for the system operation. A more complete study on the subject may be found in [2].

3 Free Market Self-regulation

It is important to define what FM self-regulation is, to avoid possible misunderstandings. It is assumed, that it is not only a process which e.g. modifies prices, production, consumption, etc. optimizing assumed narrow criteria, but self-regulation is also a spectrum of other possible FM modifications. FM self-regulation can be a mutually related set (network) of relative simple actions over certain period of time. It can be e.g.: an introduction of new technology (efficient cars in response to high petrol prices); retraining people for another job (teachers facing demographic crisis); rerouting of trade routes (1973 oil crisis and Suez Canal lock); shift of production to countries (like China) with low labor costs. Thus, FM self-regulation is perceived as any structural modification responding to disruption. FM self-regulation and disruption must be logically linked in terms of: $FM_{disruption} \xrightarrow{method} FM_{self-regulation}$. We should keep in mind that many FM processes are difficult to observe and the most suitable tool at the moment is the use of fuzzy logic. In section 5, a simple example taken from economic history will be given. This can be well modeled in our theory.

4 Implied Computations - Example

Let's give an extremely simple example of implied computations on the platform of Turing Machine (TM) - see Fig. 1. There are two different views on TM tape, which represents an extremely simple FM: a business view $V_{business}$ where *value* is the driving force for control and computational view V_{comp} where unary addition is the goal. The tape is observed by the set of heads, representing agents, looking for elements with high *value*. They act in a parallel, independent way, to reflect the real world, where set of agents observe the FM; they infer and do business.

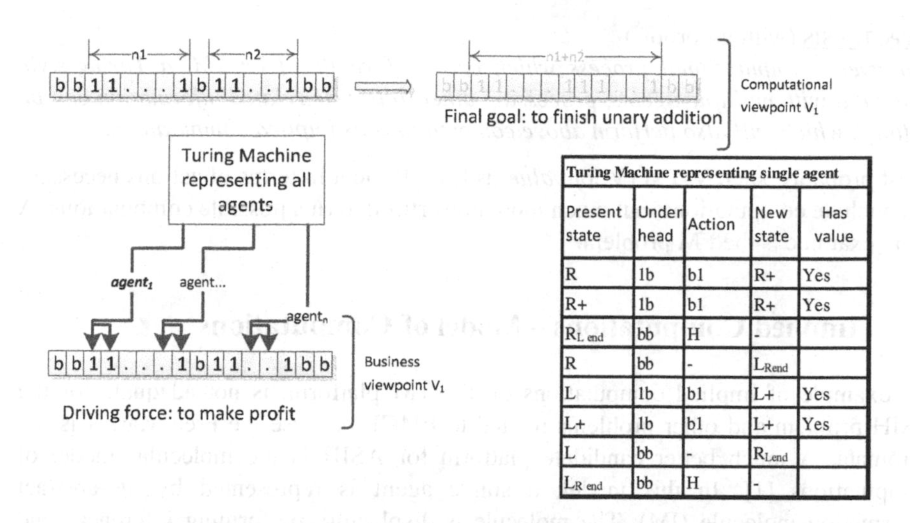

Turing Machine representing single agent				
Present state	Under head	Action	New state	Has value
R	1b	b1	R+	Yes
R+	1b	b1	R+	Yes
$R_{L\ end}$	bb	H	-	
R	bb	-	L_{Rend}	
L	1b	b1	L+	Yes
L+	1b	b1	L+	Yes
L	bb	-	R_{Lend}	
$L_{R\ end}$	bb	H	-	

Fig. 1. Implied computations in automata based on Turing Machine concept

The following set of rules define activity of the single agents:

1. TM Head representing a single agent, observes two neighboring cells;
2. Agents are initialized randomly along the tape, in such a way, that their observation areas do not overlap. Only ends (edges) of the tape are defined;
3. Agents can move left or right, only one cell per simulation step. If a given cell is observed by another agent, the agent will jump over;
4. Processing takes place when an agent discovers sequence `1b` which has *value*. He will replace it by sequence `b1` i.e. move `1` to right;
5. Halt state `H` will happen when any agent reaches right end, traveling from the left end of the list in value seeking mode and finally cannot make any further operation. In such a case, the agent will publicly announce that there is no more chance for business. The agent will do this by posting `H` state to inform all. All simulation process will be halted. Similar scenario can happen, when traveling to opposite direction;
6. A single agent can be in mental state: `R` (traveling to the Right and looking for *value*), `L` (traveling to the Left and looking for *value*), `R+` (traveling to the Right after successful action and looking for the next action to be done), `L+` (traveling to the Left after successful action and looking for the next action to be done), R_{Lend} , L_{Rend} when traveling back (return) along the list, without any successful action;
7. When reaching leftmost/rightmost sequence of symbols bb ending the tape, the mental state of the agent: `R+`/`L+` will be reset to `L`/`R` respectively, i.e. "go back";

Conclusion: business agents doing their own business unconsciously and "automatically" in parallel, have also generated a sum of two unary coded numbers. Thus now, if this sum feeds back to TM control, it can be said that FMIH exists in TM.

HYPOTHESIS (without proof[5])

For every computational process which can be formalized by TM, a Turing-style automata with multiple agents, processing other, business related information can be defined, which will also perform above computations as implied computations.

Most probably, such V-TM (V for *value*) is less efficient in terms of actions necessary to conclude computations, but much more powerful in terms possible computations. A good example is the FM problem.

5 Implied Computations – Model of Computations

An example of implied computations on the TM platform, is not adequate for the ASIH problem and other problems related to FMCI, because the Free Market is not automata. A much better candidate platform for ASIH is the molecular model of computations [1]. In this model, a single agent is represented by an abstract information molecule (IM). The molecule is displacing, performing inferences and business actions in the space of a FM; which is considered (modeled) as computational space (CS) for all agents acting and interacting there. A full theory explaining the hierarchy of information molecule, how they can be nested and how they interact with computational space (CS) can be found in [11]. The term "information molecule" is a much more suitable and safe term, in comparison to "agent", as it allows us to describe with the help of information molecule elements[6] (such as "transaction") the transfer of value, etc. Thinking in terms of IM and CS we should abstract from geometrical space of our world where the FM is nested, because today most business activity is done virtually. Thus a more abstract space with only elements of 4D geometry are necessary for the ASIH. Displacements and interactions of molecules representing agents are driven not by external control, but by *value* processing (business) categorical imperative; toward the increase of *value*. Molecules representing FM objects can also be carrying and processing other facts, rules and goals belonging to threads of implied computations, but without consciousness about their real importance and destination. This creates ASIH - not related (in short term perspective) to present business activity of the agent.

Let's consider simple example:

1. Let there be given FM with (live) agents. They buy, sell, looking for best price and highest profit, etc. – i.e. "doing business as usual". Every new item they encounter during their activity, even if not related to their mainstream business, is analyzed whether it can be used as a component to build inference chain, which will describe any new, future business. Thus, *value* must be assigned to this item and combinatorial effort (analysis) must be done whether on the basis of currently available facts, rules and goals any future business can be planned;

[5] Under research investigation.

[6] This description phenomenon exists also in physics where the graviton is a hypothetical particle that mediates the force of gravitation in quantum field theory. http://en.wikipedia.org/wiki/Graviton

2. Every contact with other agent results with mutual acquiring of knowledge; what they have and want and what *values* they have assigned to items they have or want;
3. Let's assume also that the FM is in a dynamically stable state i.e. prices are stable, buyers and sellers are the same, etc. but agents are vigilantly looking for any new opportunity to make business, if possible;
4. Let's assume that following resources arc available, but they are not object of interest (inferences) because *values* assigned to them are low:
   (a) Some agents own *tinder*. They are willing to sell it, but nobody wants it;
   (b) Some agents own *fire striker*[7] . They are willing to sell it, as above;
   (c) Some agents own *flint*. As above;
   (d) Some agents own *fuel* e.g. dried wood. As above;
   (e) Some agents somewhere/somehow have observed and remembered that striking flint with the striker makes sparks, which in the presence of the tinder, makes temporary fire. They can be considered by other as specialists to be employed, to generate temporary fire. Value assigned by owners of this skill to it, reflects a pay (salary, cost of service) they want to get to service this skill;
   (f) Some agents have seen and remembered that temporary fire caused by e.g. lighting in the presence of fuel will convert into permanent fire. Moreover, they have remembered the feeling of heat. They can be employed as specialists for managing fire. The value assigned to this skill, reflects a pay they want to get;

In the beginning items a) – f) are available, but with very low *value*, thus are not processed by inference engines when thinking about a business. Predicate calculus expressions describing logically the above situation, have the following form:

$$\left\{\begin{array}{l} \langle tinder. \rangle^{value=low}, \quad \langle fire\_striker. \rangle^{value=low}, \quad \langle flint. \rangle^{value=low}, \quad \langle fuel. \rangle^{value=low} \\ \langle tinder \wedge fire\_striker \wedge \text{flint} \xrightarrow{\text{making small fire}} temporary\_fire. \rangle^{value=low}; \\ \langle temporary\_fire \wedge fuel \xrightarrow{making\_permanent\_fire} permanent\_fire. \rangle^{value=low} \\ \langle permanent\_fire \xrightarrow{heating} heat. \rangle^{value=low} \end{array}\right\}$$

The above listed facts and rules will be initially spread around society of agents in a random way. After a certain time, on the basis of rendezvous when doing basic business, agents will gather knowledge / memory, of who has what components. Suppose that a disturbance has occurred in the FM. Let it be a global cold snap. As a result, in consciousness of some (or all) agents, a strong demand for heat will emerge. It can be coded as: $\langle ?\,heat. \rangle^{value=high}$ because all agents will be willing to buy it. As mentioned, parameter *value* labeling logical expressions is driving the inference processes; which requires meta-rules for processing this parameter. Some are given below. Thus, the emergence of expression *?heat.* with *value=high* will gradually propagate the increase of *value* to *high* for all components required for this chain of inference.

---

[7] `http://en.wikipedia.org/wiki/Fire_striker`

meta-rule1:

$$\frac{\langle \text{any logical expression } X\rangle^{value=low} \wedge \langle \text{somebody wants } X\ \rangle}{\langle X\rangle^{value=high}}$$

meta-rule2:

$$\frac{\langle\{\text{any inference with components } \langle a_1\rangle^{value=low},...,\langle a_n\rangle^{value=low}\}\text{conclusion }\langle X\rangle^{value=high}\rangle}{\langle a_1\rangle^{value=high},...,\langle a_n\rangle^{value=high}}$$

Example:

$$\frac{\left\{\begin{array}{l}\langle tinder.\rangle^{value=low},\langle fire\_striker.\rangle^{value=low},\\ \langle tinder \wedge fire\_striker \wedge \text{flint} \xrightarrow{\text{making small fire}} temporary\_fire.\rangle^{value=low}\end{array}\right\} \quad \text{conclusion } \langle temporary\_fire.\rangle^{value=high}}{\langle tinder.\rangle^{value=high},\langle fire\_striker.\rangle^{value=high},\ \langle tinder \wedge fire\_striker \wedge \text{flint} \xrightarrow{\text{making small fire}} temporary\_fire.\rangle^{value=high}}$$

As a result, these components will be on the top of business agenda of agents in terms of acquiring them and in terms of inferences based on these components.

6 Example of FM Self-regulation

The FM self-regulation is a natural consequence of the presented theory. It can even be said that it is "built-in". Probably the simplest example, reflecting real economic situation with ASIH, which can be relatively easily modeled with proposed model, is to model relations between trade in the 15th century and great geographical discoveries (Vasco da Gama, Columbus, Magellan). In early 15th century trade, transport of goods and trade was mainly concentrated in the area of the Mediterranean Sea, Baltic Sea and the English Channel. Mediterranean Sea trade was stable and was based on trading, logistic and military power of Italian republics (Genoa, Venetia), and Arabs who were great traders[8]. This FM can be modeled with simple model of agents buying, selling and transporting e.g. oriental spices from Far East to Middle East and to Western Europe. A visualization can be done through GIS[9]-style computer graphics presentation of transactions and transportation over the map of Far East, Middle East and the Mediterranean Sea. After initializing and start of simulation process, after a certain period of time, this dynamic map of trade will stabilize, like it happened in the real world in the past. Shipyards working at that time for this market, were producing only short-range ships. Similarly for the North Sea and the Baltic Sea. Ships used in this period for Mediterranean Sea were galley-style with Latin sail. Technology of ocean-going ships (galleon) was slowly emerging that time, but demand for big and expensive galleons was low. However, later on in the discussed

[8] For Arabs (nomads) traveling and trade was fundamental element of their culture. Source: `http://biznes.pwn.pl/index.php?module=haslo&id=3897007`

[9] GIS – Geographic Information Systems.

period of time, Turks[10] invading Anatolia and Middle East, have imposed high taxes on spice trade, which resulted in a large increase of prices. This implied very strong logic demand for a solution of this FM disturbance. This logic demand was strongest at the end of the supply chain i.e. in Spain and Portugal where demand for spices was very high because of oriental-style cuisine inherited from Moors. Suppose now, that we describe this situation with logical expressions, with overriding goal to reach India, to import spices. Suppose next, that the mechanism of random search around geographical map available that time will be build-in into the model, with implied update of this map. Such process will demand bigger and tougher ships which will provide positive feedback in such way, that during each next simulation cycle, braver and longer distance expeditions will be performed. As a result, in the simulation model after certain period of time:

- America will be discovered;
- Sea route around the Cape of Good Hope will be discovered;
- Magellan's trip around the globe will be done;
- In general, agents acting in this simulation model will gradually build almost complete map of Earth and its continents.

The symptom of Free Market self-regulation will be easily visible as relative rapid change of map of transactions and trips related to spice trade. The network of transactions in the Mediterranean Sea will be significantly reduced in favor of trips to America and around Cape of Good Hope to India.

7 Conclusion

We expect to have a ready simulation model in September, illustrating the presented theory and a working example from section 5. The model is built in Java, on Eclipse platform, using Mason multiagent simulation tool with GeoMason geospatial support for MASON. Special emphasis will be placed on computer graphics animated part of the presentation[11]. It will be demonstrated that such a system is in fact a computer, because with the input of proper clauses is able to make other calculations, e.g. addition-style, find max/min value calculations. They are done in parallel to mainstream inference processes related to standard business activity. Secondary school textbooks define *Economics as a study of human behavior*. Today economics as a science describes the behavior and dynamics of this highly dynamical world in terms of money, production, consumption, reserves, etc. and derivatives of these terms like e.g. inflation, unemployment, GDP per capita, increase or decrease of GDP[12], etc. This paper suggests, that businessmen, financiers, banks, market analysts, etc. thinking about future profit should change:

[10] Ottoman Empire was of military administration nature, not based on trade and production.
[11] T. Szuba is head of Computer Graphics Lab at AGH University.
[12] Gross Domestic Product.

- The way of thinking about the market, from a point of view with money in focus, to a point of view concentrating on “Free Market Collective Intelligence”. Financial profits or loses should be derivatives of this factor;
- Tools used. The simulation model we built is perhaps a progenitor of future systems. Such market analysis and prediction tools will require a much more general education and creative imagination from market analysts, to allow them to properly program and analyze simulations;
- It is expected that much stronger computers will be necessary, comparing to what banks use currently. Authors personally believe that computer systems comparable to those used for earthquake prediction (seriously!), will be necessary e.g. the system built for Japanese government's initiative "Earth Simulator Project", http://www.jamstec.go.jp/es/en/system/system.html.

References

1. Adleman, L.M.: Molecular computation of solutions to combinatorial problems. Science 266(11) (1994)
2. Buda, R.: Market Exchange Modeling Experiment, Simulation Algorithms, and Theoretical Analysis, MPRA paper no 4196. University Library of Munich, Germany (1999)
3. Eberbach, E.: The $-calculus process algebra for problem solving: A paradigmatic shift in handling hard computational problems, http://www.sciencedirect.com/science/article/pii/S0304397507003192
4. Jorion, P.: Adam Smiths Invisible Hand Revisited. An Agent-Based simulation of the New York Stock Exchange (2005), http://www.pauljorion.com/index-page-7.html
5. Peters, E.E.: Fractal Market Analysis - Applying Chaos Theory to Investment and Economics. Wiley (1994)
6. Schnabl, H.: Close Eye on the Invisible Hand. Journal of Evolutionary Economics 6, 261–280 (1996)
7. Seppercher, P.: Un modele macroeconomique multi-agents avec mon-naie endogene. Universite de la Mediterranee - Aix-Marseille II (2009)
8. Smith, A.: An Inquiry into the Nature and Causes of the Wealth of Nations. 1776
9. Skrzynski, P., Szuba, T., Szydło, S.: Collective intelligence approach to measuring invisible hand of the market. In: Jędrzejowicz, P., Nguyen, N.T., Hoang, K. (eds.) ICCCI 2011, Part II. LNCS (LNAI), vol. 6923, pp. 435–444. Springer, Heidelberg (2011)
10. Skrzynski, P.: Using Collective Intelligence Theory to Analysis of Invisible Hand of the Market Paradigm, Ph. D. dissertation, AGH University of Science and Technology Press KU 0450, Krakow (2012)
11. Szuba, T.: Computational Collective Intelligence. Wiley Series on Parallel and Distributed Computing (February 2001)
12. Szuba, T., Szydło, S., Skrzynski, P.: Computational model of collective intelligence for meta-level analysis and prediction of free or quasi-free market economy. In: XIII Int. Conf. on Corporate Governance - Theory and Practice, Poland, Cracow, November 17-18 (2011)
13. Ygge, F.: Market-Oriented Programming and its Application to Power Load Management, Ph. D. dissertation, Lund University (1998)

An Approach to Norms Assimilation in Normative Multi-agent Systems

Moamin A. Mahmoud[1], Mohd Sharifuddin Ahmad[2], Azhana Ahmad[2], Mohd Zaliman Mohd Yusoff[2], Aida Mustapha[3], and Nurzeatul Hamimah Abdul Hamid[4]

[1] College of Graduate Studies, Universiti Tenaga Nasional, 43000 Kajang, Selangor, Malaysia
[2] College of Information Technology, Universiti Tenaga Nasional, 43000 Kajang, Selangor, Malaysia
[3] Faculty of Computer Science & Information Technology
[4] Universiti Putra Malaysia, Serdang, Selangor, Malaysia
Faculty of Computer and Mathematical Sciences, Universiti Teknologi MARA

moamin84@gmail.com, {sharif,azhana,zaliman}@uniten.edu.my, aida@fsktm.upm.edu.my, nurzeatul@tmsk.uitm.edu.my

Abstract. Empirical researches on norms have become the subject of interest for many researchers in norms and normative systems. However, research in norms assimilation seems to be lacking in concept and theory within this research domain. In this paper, we propose a norms assimilation theory and implement such assimilation by exploiting an agent's internal and external beliefs. Internal belief represents the agent's ability while the external belief refers to the assimilation cost with a specific social group or society. From the agent's beliefs of its ability and assimilation cost, it is able to decide whether to accept or reject assimilation with a particular social group. This paper builds an approach to norms assimilation and analyzes the cases for an agent to decide to assimilate with a social group's norms.

Keywords: Intelligent software agents, norms, normative systems, norms assimilation.

1 Introduction

Norms and normative systems have become interesting research areas because they enhance the predictability of a society [1, 2, 3]. The concepts of norms are used to define the behaviors of agents in multi-agent systems. Basically, norms represent acceptable behaviors for a population of a natural or artificial society. In general, they are known as rules indicating actions that are expected to pursue or avoid, which are obligatory, prohibitive or permissive based on a specific set of facts [4].

While empirical research on norms have been the subject of interest for many researchers, norms assimilation has not been discussed formally. Crudely put, norms assimilation is the process of joining and abiding by the rules and norms of a social

A. Bădică et al. (eds.), *Recent Developments in Computational Collective Intelligence*, Studies in Computational Intelligence 513,
DOI: 10.1007/978-3-319-01787-7_2, © Springer International Publishing Switzerland 2014

group. Eguia [5] defined assimilation as the process in which agents embrace new social norms, habits and customs, which is costly, but offers greater opportunities. The problems of norms assimilation are attributed by the ability and capacity of an agent to assimilate in a heterogeneous society, which entails a number of social groups that have different normative protocols (in compliance and violation) and the motivation required for the agent to assimilate with a better-off group [5].

Most empirical research on norms are concerned with norms creation and enforcement [6, 7, 8]; norms spreading and internalization [9, 10]; norms emergence in agent societies [11, 12]; emotional agent architecture [13, 3]; recognition [14, 15, 16]; norms adaptation [17, 18]; and norms identification [19, 20]; normative framework architecture [3]. A literature search within the domain of norms and normative systems does not seem to produce a substantial number of research papers that discuss the empirical approach to norms assimilation. The papers only discuss the meaning of the word 'assimilation' without building any concrete concept about it [21, 22, 23]. However, the concept of assimilation has been discussed in the domain of social sciences deliberating on the assimilation cost between two social groups concerning the difference in assimilation costs between better-off and worse-off groups or between minority and majority groups [5, 24, 25].

This paper is an extension to our work in Norms Detection and Assimilation [26], in which a visitor agent detects the norms of a group of local agents with the intention to assimilate with the local agents in pursuit of its goal. The objective of this study is to build an approach for norms assimilation and analyze the cases for an agent to decide to assimilate with a social group. In this paper, we develop a norms assimilation approach based on an agent's internal belief about its ability and its external belief about the assimilation cost of a number of social groups. The assimilation cost consists of maximum and threshold costs. The maximum cost is attributed by ideal behaviours, i.e., if an agent assimilates them in a social group, it becomes optimal. The threshold cost is the minimum cost to go through assimilation. From the agent's belief about the ability and costs, it is able to decide whether to accept or reject the assimilation with a specific social group or it may join another group based on its beliefs.

The next section dwells upon the related work on norms assimilation. Section 3 details out the assimilation approach. Section 4 discusses the output of assimilation. In Section 5, we present the preliminary results and Section 6 concludes the paper.

2 Related Work on Norms Assimilation

In the assimilation of normative multi-agent systems, the related research in this area are very few, and most of them do not formally discuss the issue [21, 22, 23]. However, some research in social studies have discussed on the subject of heterogeneous society with a number of social groups possessing competing social norms [5, 24, 25]. According to Eguia [5], for an agent to be able to assimilate with a new social group, it is required to embrace the group's social norms, which is costly, but offers superior opportunities. The cost of assimilation depends on the attitude of the better-off group that the migrant or immigrant wants to join.

Eguia [5] proposed a theory in norms assimilation, in which there are two types of agents; advantaged agents and disadvantaged agents and there are also two types of groups; better-off group and worse-off group. Any disadvantaged agents can choose to join the worse-off group without cost, or it can learn to enhance its ability to be able to assimilate with the better-off group but the enhancement is costly. He found that advantaged agents optimally screen those who want to assimilate by imposing a difficult assimilation process such that the agents who assimilate are those whose ability is sufficiently high so that they generate a positive externality of the group. Members of the relatively worse-off group face an incentive to adopt the social norms of the better-off group and assimilate into it. The cost of assimilation is endogenous and chosen by the better-off group to screen those who wish to assimilate. Eguia [5] shows that, in equilibrium, only high type advantaged agents who generate positive externalities to the members of the better-off group will assimilate. In another work by Konya [24], he examined the population dynamics of heterogeneous society that has two ethnic groups, a minority and a majority. Minority members have the choice whether to assimilate with the majority. He found that, the long-run equilibrium is efficient in various cases. However, assimilation can be too slow, when minority members do not take into account the positive external effect of their decision on the majority. In general, the results show that small minorities are likely to assimilate, whereas large ones are not [24].

The literature discussed several empirical researches in norms internalization [10, 27]. To avoid confusion between norms assimilation and norms internalization, we discuss the difference in this section. When we say that an agent is able to assimilate specific norms, we mean that the agent has already detected the norms and is going through internalization process to establish those norms. According to Verhagen [27], internalization is the process in which agents integrate information into their cognitive structure [10], while assimilation, as defined earlier, is the process in which agents embrace new social norms, habits and customs, which is costly, but offers greater opportunities [5]. For example, let's assume an agent, α, would like to join a social group, σ. The social group, σ, imposes that for any agent, which would like to join the group, it has to be able to assimilate the set of norms (υ, ι, χ). After agent α detects the required norms, it decides based on two situations:

- When agent α is able to assimilate the set of norms (υ, ι, χ), then it has the choice to join with low cost of assimilation.
- When agent α is not able to assimilate all the norms, but it is able to assimilate (υ, ι) only, then there are two cases of decision:
  - o If the agent accepts the cost of assimilation, it needs to go through the internalization process to establish the norm (χ). Subsequently, the agent is able to join the social group.
  - o If the agent rejects the cost of assimilation, the social group rejects any attempts of this agent to assimilate.

From the above, we notice that the assimilation is equivalent to the internalization of non-internalized norms.

3 The Assimilation Approach

3.1 The Assimilation Model

We develop an assimilation model based on an agent's internal belief about its ability and its external belief about the cost of assimilation with a specific social group. As shown in Figure 1, while the agent has its internal belief about its ability it detects the various social groups' norms and calculates the cost of assimilation for each group. Based on its ability and the cost, it decides which social groups to assimilate with.

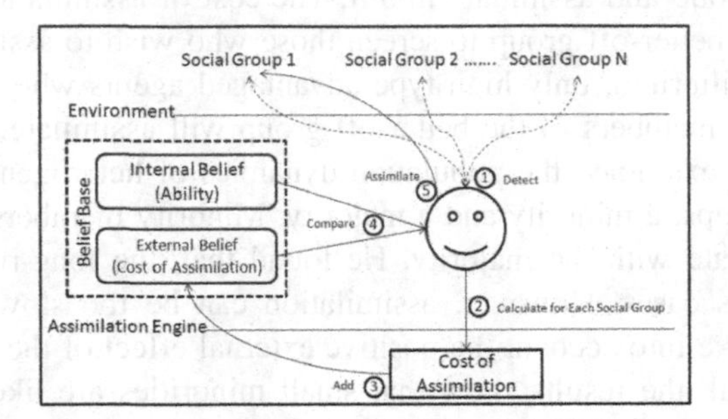

Fig. 1. The Assimilation Model

From Figure 1, there is an agent, and a number of social groups. (1) The agent first detects the norms of these groups. Then, (2) it calculates the cost of assimilation with each social group based on their practiced norms. (3) The agent then adds the various costs of these groups to its external belief. (4) From its internal belief that contains its ability and its external belief that contains the cost of assimilation with these social groups, (5) it decides which group to assimilate with, and in Figure 1, we assume that the agent chooses the Social Group 1.

3.2 The Influential Elements on Assimilation

According to Eguia [5], the decision to assimilate is influenced by two elements, one belongs to the agent's internal belief, which is the ability, and the other belongs to its external belief, which is the assimilation cost for specific social group. The assimilation cost consists of the maximum and threshold costs. We define these elements as follows:

Definition 1: The ability, β, it is the competence and qualification of an agent to assimilate the norms. For an agent to join a social group, it must be able to assimilate their norms.

Definition 2: The Assimilation Cost, Σ, is the total effort and expenses incurred by an agent to assimilate with a social group. It consists of two types: the maximum and the threshold assimilation costs.

Definition 3: The Maximum Assimilation Cost, μ, is the highest cost imposed by a social group for assimilation. Any agent which is able to meet this cost is considered as optimal agents. An optimal agent is an agent, which has the competence to practice all required norms of a social group.

Definition 4: The Threshold Assimilation Cost, τ, is the minimum acceptable cost to assimilate with a specific social group. The threshold cost is different from one social group to another. Based on these elements, agents and social groups can decide to accept or reject any assimilation attempts with the social groups. However, there are three cases to consider, two of which are favourable to the agent for assimilation. In the other case, the agent is not welcome to assimilate with the social group.

First Case: The first case is shown in Figure 2, when the value of the agent's ability is greater than the value of the threshold assimilation cost. Suppose there is an agent, α, which has the ability value, θ_β, and there is a social group, σ which imposes the maximum assimilation cost value, θ_μ, and the threshold assimilation cost value, θ_τ. Let λ, shown in Figure 2, represents the reference vector between τ and β. When $\theta_\beta - \theta_\tau > \lambda$, we say the agent α can assimilate σ,

$\theta_\beta - \theta_\tau > \lambda \Rightarrow$ can (assimilate (α, σ)) Case 1

We choose the term *can* because the agent's ability is greater than the threshold assimilation cost, which means that the agent, α, is provably competent and qualify to assimilate with the group. For example, if the minimum Cumulative Grade Point Average (CGPA) (θ_τ) to join σ is 3.0, and the imposed maximum assimilation cost value (θ_μ) is 3.5, and that agent's (α) CGPA (θ_β) is 3.3, then α's ability is provably greater than the threshold assimilation cost value and it is definitely recommended to join the group. In Figure 2, we can see that the vector δ is the actual ability value that is equal to τ with reference to λ,

If θ_δ is the value of δ then, the value between δ and λ equals the value between τ and λ.

$$\theta_\delta = \theta_\tau \tag{1}$$

The value between β and δ equals the extra ability, θ_ρ, that the agent has.

$$\theta_\beta - \theta_\delta = \theta_\rho \tag{2}$$

From 1 and 2,

$$\theta_\beta - \theta_\tau = \theta_\rho \tag{3}$$

From 3, we can determine the value of β and τ for this case,

$\theta_\beta = \theta_\tau + \theta_\rho$, $\theta_\tau = \theta_\beta - \theta_\rho$

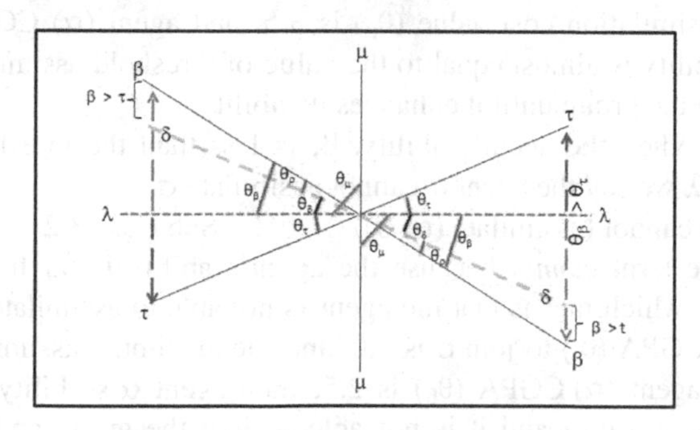

Fig. 2. Ability More than Threshold

Second Case: The second case is as shown in Figure 3, when the value of the agent's ability is equal to the value of threshold assimilation cost. When $\theta_\beta - \theta_\tau = \lambda$, we say the agent α could assimilate σ,

$\theta_\beta - \theta_\tau = \lambda \Rightarrow$ could (assimilate (α,σ)) Case 2

We choose the term *could* because the agent's ability is equal to the threshold assimilation cost, which means that the agent is at a critical point of its ability to assimilate with the group. In our example, if the minimum CGPA (θ_τ) to join σ is 3.0 and the maximum assimilation cost value (θ_μ) is 3.5, and agent (α) CGPA (θ_β) is 3.0, then agent α's ability is equal to the threshold assimilation cost. In this case, it is able to join the group but it should enhance its ability. From Figure 3, the value of β equals the value of τ, with reference to λ.

$\theta_\beta - \theta_\tau = \lambda$

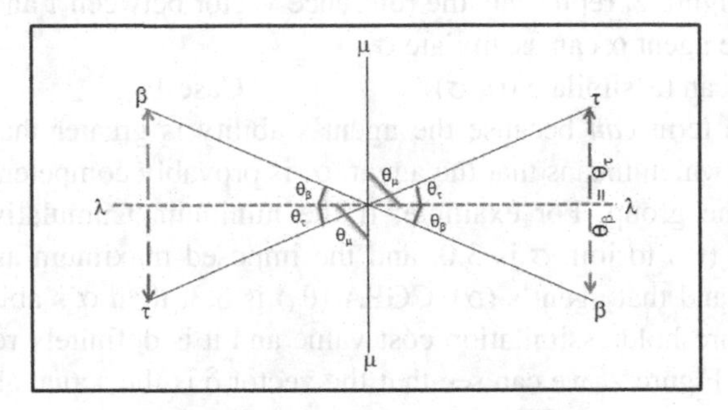

Fig. 3. Ability Equals Threshold

Third Case: The third case is shown in Figure 4, when the value of the threshold assimilation cost is greater than the agent's ability. In this case, there are two sub cases.

Sub Case 1: When the agent's ability, β, is almost equal the threshold assimilation cost, τ, $\theta_\tau - \theta_\beta \cong \lambda$ we say the agent α could not assimilate σ,

$\theta_\tau - \theta_\beta \cong \lambda \Rightarrow$ could not (assimilate (α, σ)) Sub Case 3.1

We choose the term *could not* because the agent's ability is almost equal to the threshold assimilation cost, which means that the agent is not able to assimilate within this group until it enhances its ability. If the minimum CGPA (θ_τ) to join σ is 3.0, and the maximum assimilation cost value (θ_μ) is 3.5, and agent (α) CGPA (θ_β) is 2.98, then agent α's ability is almost equal to the value of threshold assimilation cost and it is not able to join the group until it enhances its ability.

Sub Case 2: When the agent's ability, β, is less than the threshold assimilation cost, τ, $\theta_\tau - \theta_\beta > \lambda$ we say the agent α cannot assimilate σ,

$\theta_\tau - \theta_\beta > \lambda \Rightarrow$ cannot (assimilate (α, σ)) Sub Case 3.2

We choose the term *cannot* because the agent's ability is far from the threshold assimilation cost, which means that the agent is not able to assimilate with the group. If the minimum CGPA (θ_τ) to join σ is 3.0, and the maximum assimilation cost value (θ_μ) is 3.5, and agent (α) CGPA (θ_β) is 2.5, then agent α's ability is less than the threshold assimilation cost and it is not able to join the group and assimilate their norms. From figure 4, the vector δ is the actual ability value that is less than τ, with reference to λ. If θ_δ is the value of δ then,

The value between δ and λ equals the value between β and λ, i.e.,

$$\theta_\delta = \theta_\beta \tag{4}$$

The value between τ and δ equals the extra cost, θ_ρ, which the agent does not have.

$$\theta_\tau - \theta_\delta = \theta_\rho \tag{5}$$

From 4 and 5,

$$\theta_\tau - \theta_\beta = \theta_\rho \tag{6}$$

From 6 we can determine the value of β and τ for this case,
$\theta_\beta = \theta_\tau - \theta_\rho$, $\theta_\tau = \theta_\beta + \theta_\rho$

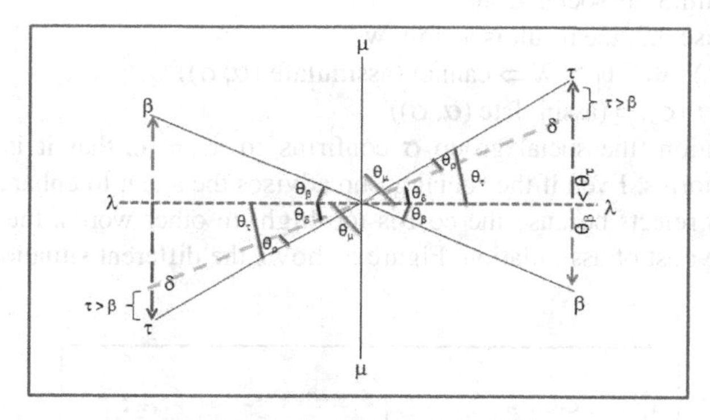

Fig. 4. Threshold More than Ability

3.3 Discussion

In this section, we discuss the assimilation cases and the result for each case from a social group and from an agent itself, from above, the cases as follow,

$\theta_\beta - \theta_\tau > \lambda \Rightarrow$ can (assimilate (α, σ))	Case 1
$\theta_\beta - \theta_\tau = \lambda \Rightarrow$ could (assimilate (α,σ))	Case 2
$\theta_\tau - \theta_\beta \cong \lambda \Rightarrow$ could not (assimilate (α, σ))	Sub Case 3.1
$\theta_\tau - \theta_\beta < \lambda \Rightarrow$ cannot (assimilate(α, σ))	Sub Case 3.2

If there is an agent, α, and there is a social group, σ,

- In Case 1, the result,

  Ask (α): $\theta_\beta - \theta_\tau > \lambda \Rightarrow$ can (assimilate (α, σ))
  Answer (σ): (assimilate (α, σ))

  In this situation, the social group σ confirms to agent α that it is able to assimilate the norms.

- In Case 2, the accepted result is as follows,

  Ask (α): $\theta_\beta - \theta_\tau = \lambda \Rightarrow$ could (assimilate (α, σ))
  Answer (σ): should (enhance (α, β) $\rightarrow \beta \geq \lambda$) $\otimes$ (assimilate (α, σ))

In this situation, the social group σ suggests to agent α to enhance its ability before it is allowed to assimilate the norms. In this case, the agent, as an autonomous entity, has the choice to accept or reject the suggestion, and if it rejects, the agent is still able to assimilate within this social group, but if there is any slack in its performance, the social group penalizes the agent and revokes its assimilation with the group. However, the agent is recommended to be in the safe zone.

- In Sub Case 3.1, the result,

  Ask (α): $\theta_\tau - \theta_\beta \cong \lambda \Rightarrow$ could not (assimilate (α, σ))

  Answer (σ): must (enhance (α, β) $\rightarrow \beta \geq \lambda$) $\wedge$ (assimilate (α, σ))

In this situation, the social group σ sets a condition to agent α that it must enhance its ability before it is allowed to assimilate the norms. In other words, the assimilation is conditioned with ability improvement. In this case, the agent, as an autonomous entity, has the choice to accept or reject the condition, but if it rejects, it is not allowed to assimilate within the social group.

- In Sub Case 3.2, the result is as follows,

  Ask (α): $\theta_\tau - \theta_\beta > \lambda \Rightarrow$ cannot (assimilate (α, σ))

  Answer (σ): $\neg$ (assimilate (α, σ))

In this situation, the social group σ confirms to agent α that it is not able to assimilate the norms. Even if the social group advises the agent to enhance its ability, the agent often rejects because the cost is too high. In other words, the agent is not able to bear the cost of assimilation. Figure 5 shows the different situations discussed above.

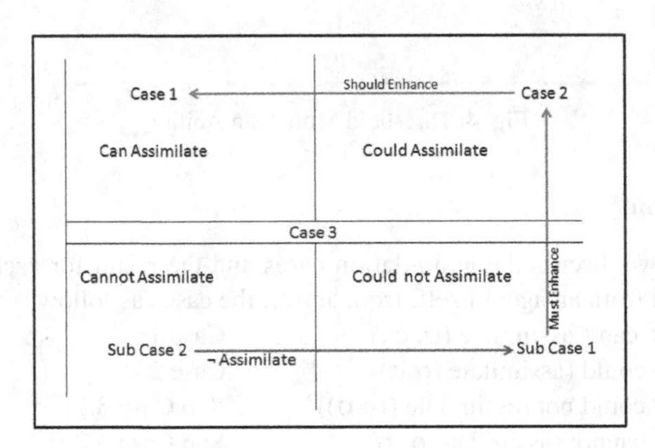

Fig. 5. Decision Cases

4 Preliminary Results

The preliminary results highlight the following points:

- An agent's decision to assimilate falls into one of the four cases, which are: can assimilate; could assimilate; could not assimilate; cannot assimilate.
- The assimilation cost is based on the value of threshold assimilation cost.

- The threshold assimilation cost values are based on a social group's compliance with their norms. High compliance increases the threshold value and subsequently increases the assimilation cost, while low compliance reduces the assimilation cost for the same reason.
- When an agent's ability is more than the threshold cost, the agent is recommended to join and proceed with the assimilation.
- When an agent's ability is equal to the threshold cost, the agent is advised to enhance its ability before going through the assimilation process.
- When an agent's ability is almost equal to the threshold cost, the agent is required to enhance its ability before going through the assimilation.
- When an agent's ability is less than the threshold cost, the agent is not able to assimilate and it is not welcomed by a social group.

5 Conclusion and Further Work

In this paper, we present our research on norms assimilation in a heterogeneous society where there are a number of social groups observing similar norms. Any agent, which would like to join a social group, has to be able to assimilate their norms. The suggested assimilation approach is based on the internal and external agent's beliefs. The internal belief represents the agent's ability and the external belief represents the assimilation cost of a social group. The preliminary results show that an agent's decision (can assimilate; could assimilate; could not assimilate; cannot assimilate) is based on the ability of the agent and the threshold assimilation cost of a social group. As an autonomous entity, it has the choice to accept or reject any advice or condition. The compliance with norms increases the threshold assimilation cost and attract prominent agents to assimilate and vice versa.

In our future work, we shall develop a method to calculate the values of ability, maximum and threshold assimilation costs. We then create a virtual environment of a typical restaurant scenario to study the cases of assimilation (can, could, could not, cannot) and to conduct experiments on the smoking phenomenon to demonstrate how the values increase and decrease based on the restaurant's social group compliance to its norms.

Acknowledgments. This project is sponsored by the Malaysian Ministry of Higher Education (MOHE) under the Exploratory Research Grant Scheme (ERGS) No. ERGS/f/2011/STG/ UNITEN/02/8.

References

1. Broersen, J., Dastani, M., van der Torre, L.W.N.: Resolving Conflicts between Beliefs, Obligations, Intentions, and Desires. In: Benferhat, S., Besnard, P. (eds.) ECSQARU 2001. LNCS (LNAI), vol. 2143, pp. 568–579. Springer, Heidelberg (2001)
2. Sadri, F., Stathis, K., Toni, F.: Normative KGP agents. Computational and Mathematical Organization Theory 12(2), 101–126 (2006)

3. Ahmad, A.: An Agent-Based Framework Incorporting Rules, Norms And Emotions (OP-RND-E), PhD Thesis, Universiti Tenaga Nasional (2012)
4. Gonçalves, R., Alferes, J.J.: Specifying and reasoning about normative systems in deontic logic programming. In: Proceedings of the 11th International Conference on Autonomous Agents and Multiagent Systems, AAMAS 2012 (2012)
5. Eguia, J.X.: A theory of discrimination and assimilation. New York University (2011)
6. Verhagen, H.: Norm Autonomous Agents, Stockholm: Department of System and Computer Sciences. The Royal Institute of Technology and Stockholm University (2000)
7. Boella, G., van der Torre, L.: An architecture of a normative system: counts-as conditionals, obligations and permissions. In: AAMAS 2006: Proceedings of the Fifth International Joint Conference on Autonomous Agents and Multi-agent Systems, pp. 229–231. ACM, Hakodate (2006)
8. Grossi, D., Gabbay, D., van der Torre, L.: Book Title: The Norm Implementation Problem in Normative Multi-Agent Systems. In: Specification and Verification of Multi-agent Systems, pp. 195–224. Springer, US (2010)
9. Neumann, M.: Norm Internalisation in Human and Artificial Intelligence. Journal of Artificial Societies and Social Simulation JASSS 13(1), 12 (2010)
10. Hollander, C., Wu, A.: The Current State of Normative Agent-Based Systems. Journal of Artificial Societies and Social Simulation 14(2), 6 (2011)
11. Boella, G., Torre, L.V.D., Verhagen, H.: Ten Challenges for Normative Multi-agent Systems. In: Bordini, R., et al. (eds.) Dagstuhl Seminar Proceedings 08361. Schloss Dagstuhl - Leibniz-Zentrum fuer Informatik, Dagstuhl (2008)
12. Axelrod, R.: An Evolutionary Approach to Norms. The American Political Science Review 80(4), 1095–1111 (1986)
13. Hu, J., Guan, C.: An architecture for Emotional Agent. IEEE (2006)
14. Sen, S., Airiau, S.: Emergence of norms through social learning. In: Proceedings of IJCAI 2007, pp. 1507–1512 (2007)
15. Conte, R., Paolucci, M.: Intelligent Social Learning. Journal of Artificial Societies and Social Simulation, JASSS 4(1) (2001)
16. Conte, R., Dignum, F.: From Social Monitoring to Normative Influence. Journal of Artificial Societies and Social Simulation 4(2) (2001)
17. Epstein, J.: Learning to be thoughtless: social norms and individual computing. Center on Social and Economic Dynamics. Working Paper, No. 6 (2000)
18. Campos, J., López-Sánchez, M., Esteva, M.: A Case-Based Reasoning Approach for Norm Adaptation. In: Corchado, E., Graña Romay, M., Manhaes Savio, A. (eds.) HAIS 2010, Part II. LNCS, vol. 6077, pp. 168–176. Springer, Heidelberg (2010)
19. Centeno, R., Billhardt, H.: Auto-adaptation of Open MAS through On-line Modifications of the Environment. In: Proceedings of the 10th International Conference on Advanced Agent Technology, AAMAS 2011, pp. 426–427 (2012)
20. Savarimuthu, B.T.R., Cranefield, S., Purvis, M., Purvis, M.: Obligation Norm Identification in Agent Societies. Journal of Artificial Societies and Social Simulation (2010)
21. Caire, P.: A Normative Multi-Agent Systems Approach to the Use of Conviviality for Digital Cities. In: Sichman, J.S., Padget, J., Ossowski, S., Noriega, P. (eds.) COIN 2007. LNCS (LNAI), vol. 4870, pp. 245–260. Springer, Heidelberg (2008)
22. Donetto, D., Cecconi, F.: The Emergence of Shared Representations in Complex Networks. In: Proceedings of the Social Networks and Multi-Agent Systems Symposium (SNAMAS 2009), pp. 42–44 (2009)

23. Andrighetto, G., Campennì, M., Cecconi, F., Conte, R.: The Complex Loop of Norm Emergence: a Simulation Model. In: The Second World Congress, Agent-Based Social Systems. LNCS (LNAI), vol. 7, pp. 19–35. Springer, Heidelberg (2010)
24. Konya, I.: A dynamic model of cultural assimilation. Boston College Working Papers in Economics 546 (2002)
25. Quamrul, A., Galor, O.: Cultural Assimilation. Cultural Diffusion and the Origin of the Wealth of Nations. Brown University, mimeo (2007)
26. Mahmoud, M.A., Ahmad, M.S., Ahmad, A., Mohd Yusoff, M.Z., Mustapha, A.: Norms Detection and Assimilation in Multi-agent Systems: A Conceptual Approach. In: Lukose, D., Ahmad, A.R., Suliman, A. (eds.) KTW 2011. CCIS, vol. 295, pp. 226–233. Springer, Heidelberg (2012)
27. Verhagen, H.: Simulation of the Learning of Norms. Social Science Computer Review 19(3), 296–306 (2001)

A Dynamic Measurement of Agent Autonomy in the Layered Adjustable Autonomy Model

Salama A. Mostafa[1], Mohd Sharifuddin Ahmad[1], Azhana Ahmad[1], Muthukkaruppan Annamalai[2], and Aida Mustapha[3]

[1] College of Information Technology, Universiti Tenaga Nasional, Putrajaya Campus 43000, Selangor Darul Ehsan, Malaysia
[2] Faculty of Computer and Mathematical Sciences, Universiti Teknologi MARA, Shah Alam, Selangor Darul Ehsan, Malaysia
[3] Faculty of Computer Sciences and Information Technology, Universiti Putra Malaysia, Selangor Darul Ehsan, Malaysia

semnah@yahoo.com, {sharif,azhana}@uniten.edu.my, mk@tmsk.uitm.edu.my, aida@fsktm.upm.edu.my

Abstract. In a dynamically interactive systems that contain a mix of humans' and software agents' intelligence, managing autonomy is a challenging task. Giving an agent a complete control over its autonomy is a risky practice while manually setting the agent's autonomy level is an inefficient approach. In this paper, we propose an autonomy measurement mechanism and its related formulae for the Layered Adjustable Autonomy (LAA) model. Our model provides a mechanism that optimizes autonomy distribution, consequently, enabling global control of the autonomous agents that guides or even withholds them whenever necessary. This is achieved by formulating intervention rules on the agents' decision-making capabilities through autonomy measurement criteria. Our aim is to create an autonomy model that is flexible and reliable.

Keywords: Software agent, Multi-agent system (MAS), Layered Adjustable Autonomy (LAA), Autonomy measurement attributes, Decision-making.

1 Introduction

Considerable attention to human-agent dynamic interactions and the related aspects are the trends of current academic and industrial communities [1]. One such trend, the adjustable autonomy paradigm, introduces many approaches that enable humans and agents to dynamically share control and initiative such as the Mixed-Initiative [2] and Sliding autonomy [3]. Some practical examples of agent-based systems that espouse this trend are: human-robots team [4, 5], unmanned vehicles [6, 7] and drones [8].

Autonomy is a core characteristic of an intelligent agent [9]. However, autonomy measurement is a very challenging task, since it is difficult to quantitatively gauge the agent's autonomy for its future behavior based on retrospective effect of its past behavior [8]. Nevertheless, we cannot ignore the fact that, quantitative measurement of agent's autonomy enhances its decision-making quality [10]. The measurement is also helps in tracing, analyzing and predicting the performance of the agent [11].

A. Bădică et al. (eds.), *Recent Developments in Computational Collective Intelligence*, Studies in Computational Intelligence 513,
DOI: 10.1007/978-3-319-01787-7_3,

We have proposed the Layered Adjustable Autonomy (LAA) model [12, 13] to enables the agent to autonomously perform decisions and actions. In order to test the ability of the LAA model to compute and measure the autonomy level of a decision that corresponds to specific action's achievement, we propose the following hypotheses to be investigated in this paper:

Hypothesis 1 : There exists a band in which an agent's autonomy is adjustable.
Hypothesis 2 : As new knowledge is gained by the agent, its authority tends to move in a direction that increases its autonomy.
Hypothesis 3 : The new gained knowledge and authority cause the adjustment band of autonomy to progressively reduce until a single boundary separating full-autonomy and non-autonomy regions is created.
Hypothesis 4 : Ultimately, there exists a situation of a system when each of its agent's autonomy reaches an optimum configuration.

Throughout this paper, the agents' autonomy deals with the decision and action perspectives only. Our target is to maximize an agent's autonomy through the constraints of system's dynamism and performance efficiency. We assume that the agents are situated in a distributed environment and engaged in particular tasks to be achieved. Nevertheless, the interactions between humans and other software agents and their related aspects are beyond the scope of this paper.

2 Related Work in Autonomy Measurement

In 1978, Sheridan and Verplank [6] proposed a model in which autonomy is scaled to ten-levels of Teleoperator System control based on tasks complexity. They studied different constraints that are related to unmanned marine vehicle. They demonstrated that, the manner of doing a task is the key to success in its completion. What is inferior about their model are its neglect of the dynamic state and mission context [6].

An extension to the above model is suggested by Parasuraman et al. [14] with the addition of using an agent for action automation. They also proposed ten-levels of autonomy and each level represents different decision-making capabilities which are distributed between the system and human agents. They further introduced four stages of cognitive information processing in their model which, respectively, are observing, analysing, decision-making and action selection as shown in Figure 1.

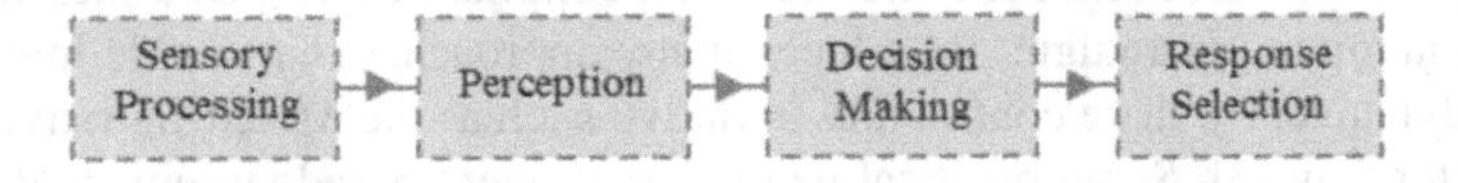

Fig. 1. Cognitive information processing by [14]

Barber and Martin [15] work is an early attempt to model an agent autonomy with regards to its decision-making aspect, quantitatively [10]. They suggested three components for autonomy representation which are the intended goal(s), G; the decision-makers of G and the relative strengths of the decision-making process, D;

and, authority constraint declared to D, C. The autonomy measurement is formed based on the tuple of the (G, D, C) variables. An agent autonomy degree is measured by weighted voting mechanism to a group of G. The model offers several advantages such as an explicit representation of autonomy and autonomy adjustment mechanism.

Brainov and Hexmoor [11] introduced a quantitative measurement for different cases of agent autonomy in MAS environment. The cases they covered are decision and action autonomy; user agent's autonomy; and distributing autonomy among a group and for different groups of agents along with the required measurements. They addressed the difficulty of modelling an algorithm that consistently matches and deploys an optimal agent or group of agents with a task to be optimally achieved.

Sliding autonomy is a recent approach in the application of adjustable autonomy. Brookshire et al. [4] proposed a sliding approach to distribute autonomy among agents. The slider is manually adjusted by a human operator to determine the responsible actor of a task. Autonomy levels are not clearly defined and the roles of the actors are determined in the action level [7]. Human-agent interactions are constrained to only switch between fully-autonomous and non-autonomous actions.

Sellner et al. [3] went further with sliding autonomy by proposing two authority modes: Mixed-Initiative Sliding Autonomy (MISA) which gives the authority to human operator to intervene and System-Initiative Sliding Autonomy (SISA) which allows the agent to decide when to request a human operator's assistance. While this approach allows the sharing of tasks between the actors, they are insufficient for critical time and dynamic systems as in the case of most Mixed-Initiative systems [1].

According to [15], a representation of autonomy must explicitly model constraints that enforce the authority of the decision-making agents. Without such enforcement, an assignment could be subverted by an agent who simply refuses to carry out the agreed upon task. The authority constraints complete the model of the decision-making by ensuring that at least one agent will execute the decisions made by the decision-making agents [16]. However, this context to autonomy is more to support the Executional type of autonomy than the Absolute or the Social autonomies [17].

The predominantly used attributes/parameters for autonomy measurement that we have reviewed in the literature are: knowledge [18, 19], authority [7, 15, 16, 18], goal-achievement [15, 20], performance [3, 7, 20], consistency [16], confidence [1, 3, 5], trust [5, 20, 21], motivation [20, 21], social integrity[17] and, dependency [2, 17, 20], efficiency [4], time [2, 3, 18] and, risk factors [5], task orientation [19] and complexity-level [4, 6, 16]. However, some of the autonomy measurement attributes are treated as indicators to other attributes' measurement (e.g. performance can be derived from complexity-level, time and goal achievement) as shown in Figure 2.

Selecting the best collection of autonomy attributes as criteria for autonomy measurement, however, is always a domain specific issue, which requires considerable analysis for optimal combination [17]. Some attributes' values are retrospectively taken by analysing past experiences [3], while some are configured based on an online situation states analysis [8]. In some advanced cases, past and/or present situations are considered in selecting the autonomy measurement attributes' values as in identifying performance related attributes [7]. In addition, some attributes are system-based [4], while others are agent-based [5]. Yet, in some advanced cases, both the agent and the system are involved in determining the attributes' values as in the authority attribute example in [3].

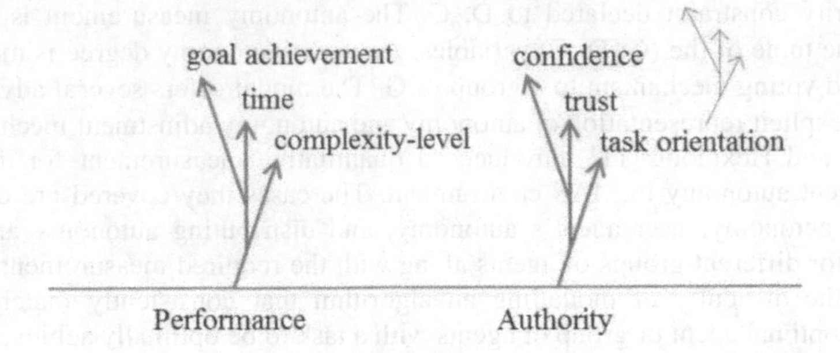

Fig. 2. Possible derivatives of autonomy measurement attributes

Nevertheless, the key indicator to autonomy adjustment is the intervention in the agent decision-making process [8, 21]. It is clear that there are two main reasons behind the interventions in agents' decisions which are the lack of its knowledge [5, 18, 19, 21, 22] and authority [7, 11, 14, 15, 16, 18] to make a particular decision. For instance, an agent with partial information and/or control over a task is more likely (by request or by force) to be intervened by others. In addition, adopting the knowledge and authority as attributes for autonomy measurement can cover different application domains.

3 Measurement of Autonomy in the LAA Model

We proposed the Layered Adjustable Autonomy (LAA) model to resolve some of the autonomy adjustment issues where software and human agents interact to achieve common goals. The LAA architecture separates the autonomy to number of layers that encapsulate an agent's decision process adjustment. Each layer is attributed to deal with actions that correspond to its autonomy level. An Autonomy Analysis Module (AAM) and Situation Awareness Module (SAM) are proposed to keep track of all the active agents to control autonomy distribution as depicted in Figure 3 [12].

Based on the agents' reaction and action selection decisions in a specific event, the system deploys a more qualified agent when the agent's threshold of autonomy level has exceeded its action of choice requirements. The more qualified agent is the one that has the knowledge (*know*) and/or authority (*can*) to handle an event using an action or set of actions with greater autonomy (i.e. agent: $know \wedge can \rightarrow decide$).

Consequently, in the LAA model, the agent's action implementation life cycle proceeds through five stages which are observing, reasoning, action selection, autonomy layer selection and action processing. To give the agent the characteristics of intelligence and speed which are the core requirements for a dynamic interactive system, the agent in the LAA is designed to works in three modes which are Deliberative, Hybrid (deliberative and reactive) and Reactive modes [23].

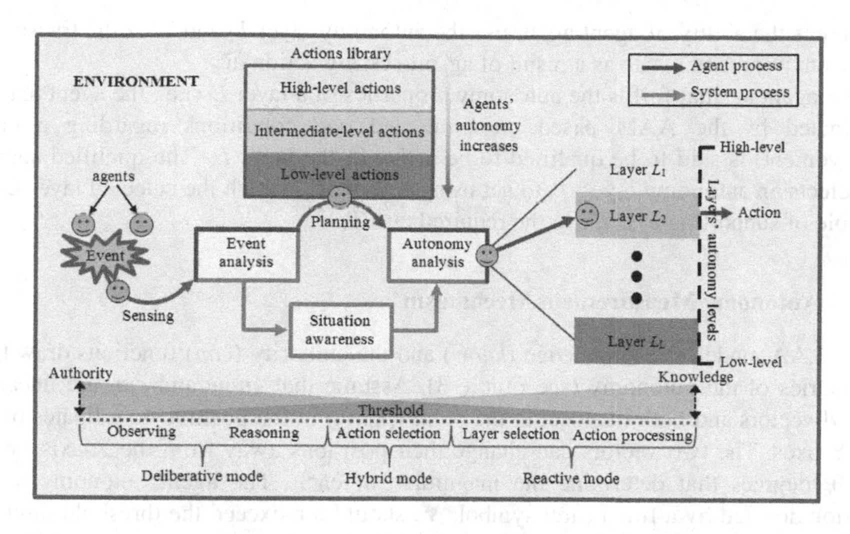

Fig. 3. General description to the LAA model

3.1 Layers and Actions

An autonomy layer can be viewed as a virtual entity that assists the deployed agent to implement actions that are directed towards optimized decision-making [12]. Each agent in the system (including human) has a set of possible actions it is capable of performing. Each layer is defined by a set of autonomy properties that are based on predefined policies and rules. The autonomy properties of one layer are different from that of the other, such as the "adjustment mechanism of actions" parameters (e.g. popping up an interface dialog box for requesting feedback from human) [13].

The actions in the LAA model are further categorized into high-, intermediate- and low-level actions, that correspond to non-autonomous, $\overline{Ac}$; semi-autonomous, $\widetilde{Ac}$; and fully-autonomous actions $\dot{A}c$.

$$Ac = \{\overline{Ac}, \widetilde{Ac}, \dot{A}c\}$$

where each of $\overline{Ac} = \{\overline{ac_1}, \overline{ac_2}, \ldots\}$, $\widetilde{Ac} = \{\widetilde{ac_1}, \widetilde{ac_2}, \ldots\}$ and $\dot{A}c = \{a\dot{c}_1, a\dot{c}_2, \ldots\}$.

Let L_L be the possible layers of the LAA model, L_l be a layer such that, $L_l \in L_L$ and the L_l has a set of possible actions $\{\overline{ac_n}, \widetilde{ac_s}, a\dot{c}_f\} \in Ac$ to be implemented in the L_l, $T_t \in T_T$ which is the proposed task to deal with an event e in the environment E by a qualified agent $ag_i \in Ag$. Let say ag_i selects action $\{\widetilde{ac_s}\}$ in event e, then ag_i action is:

$$\forall T_t \in T_T \Rightarrow \exists\, ag_i \in Ag: R^{T_t} \rightarrow \widetilde{ac_s}$$

$$\forall L_l \in L_L \Rightarrow \exists\, ag_i \in Ag: R^{L_L} \rightarrow L_l$$

$$\forall \widetilde{ac_s} \in Ac \Rightarrow \exists\, ag_i \equiv L_l: R^{\widetilde{ac_s}} \rightarrow E'$$

where Ag denotes a set of agents in the system that has responded to achieve T_t using one or more of the possible actions, R is the agent's ag_i run sequence over the e in E,

≡ denotes the ability of agent ag_i to use the autonomy layer L_l and E' is the transition of the environmental state as a result of ag_i processing action $\widetilde{ac}_s$.

The agent ag_i that fulfils the autonomy properties of a layer L_l (i.e., the agent that is nominated by the AAM based on *know* and *can* conditions regarding a task achievement) is said to be qualified to be active in the layer L_l. The qualified agent, ag_i selects an autonomy layer, L_l to act using action $\widetilde{ac}_s$, which the selected layer L_l is capable of supporting to achieve the required task T_t.

3.2 Autonomy Measurement Mechanism

In the LAA model, the knowledge (*know*) and the authority (*can*) functions draw the boundaries of the autonomy (see Figure 3). Assume that *know* and *can* are linearly related vectors and both intersect at the point $I(x_I, y_I)$ in the positive coordinates of X and Y axes. The two vectors can change their positions away from the X-axis by θ_k and θ_c degrees that determine the magnitude of each. The agent autonomy level position denoted by a filled caret symbol, ▼, should not exceed the threshold limit of *know* and *can* of the specified layer as shown in Figure 4. Each layer has different variable limit of *know* and *can* threshold which control agent's layer selection.

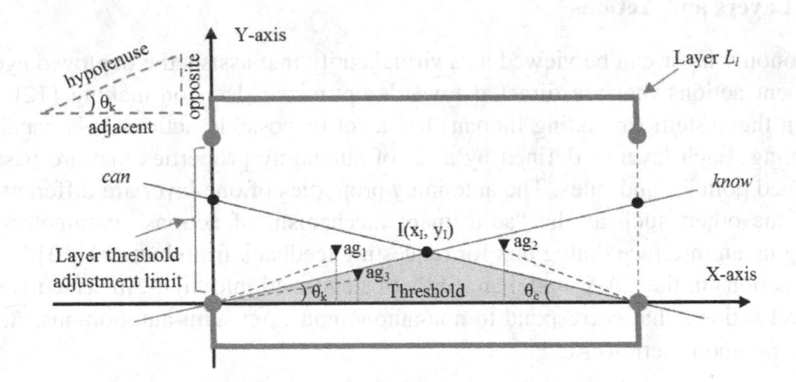

Fig. 4. Layers adjustment mechanism

In Figure 4, the X-axis represents the progress of an agent's actions and the Y-axis represents the autonomy level of the agent and the corresponding layer. To fix an adjustment model to the layer adjustment mechanism, we use cotangent trigonometric function.

$$\theta = \cot\left(\frac{y}{x}\right), \text{ if } x \neq 0 \tag{1}$$

where theta θ is the ratio of the adjacent to the opposite side of the triangle representing the progress of an agent action x, and the autonomy level y, respectively for a *know* or *can* vector of a specified layer.

Consequently, formula (1) can be used to find the theta value for each of *know* (θ_k) and *can* (θ_c) that represent: agent autonomy conditions, (i.e. agent degree of autonomy); the autonomy level of each layer (i.e. the Threshold T boundaries) and dynamically adjusting the T via increasing or decreasing *know* and *can* thetas.

If the *know* and *can* vectors are asymptotically equal with the X-axis, (i.e. y_I becomes zero) then the agent can perform the selected action without any adjustment in a fully-autonomous decision-making deployment (see Figure 4). However, this case is possibly applicable in only two situations:

- The first situation is when the action to be performed is a low-level action which means that the required thresholds of *know* and *can* are very low, thus the agent can make full-autonomous decision for any circumstances.
- The second situation is when the agent has the knowledge to successfully handle a higher-level action and hence is authorized to make full-autonomous decision regarding the action then the system adjusts θ_k and θ_c of the layer to zero.

3.3 Autonomy Distribution

Let a_i be the degree of autonomy of an agent ag_i, where $0 \leq a_i \leq 1$. Roughly, if $a_i \cong 0$, then ag_i is non-autonomous, if $a_i \cong 1$, then ag_i is fully-autonomous, otherwise, ag_i is semi-autonomous. ag_i's knowledge about event e_x is k_{e_x} ,which is calculated through ag_i's *know* function,

$$know\ (e_x) = k_{e_x},\ \text{where } 0 \leq k_{e_x} \leq 1 \tag{2}$$

ag_i's authority about handling e_x is c_{e_x}, which is calculated through the system's *can* function,

$$can\ (e_x) = c_{e_x},\ \text{where } 0 \leq c_{e_x} \leq 1 \tag{3}$$

Definition 1. The degree of an agent's autonomy during the observation and reasoning stages in an event is the degree of its knowledge about the event.

$$\forall\, e_x\ \exists\, ag_i\text{: } know\ (e_x) \Rightarrow a_i = k_{e_x} * \gamma \tag{4}$$

where γ is the intervention bias whose value is either 0 or 1. γ is calculated based on the corresponding agent performance progress by the following function composition i.e. (5).

$$\gamma = \Phi_i\ (\lambda_{e_x}\ (p\ (s, t)) \tag{5}$$

where Φ is the function of the AAM on track i of the agent ag_i and its autonomy level a_i. Φ returns a truth value based on the SAM function result, λ, which checks ag_i performance progress when handling event e_x. p is the satisfactory fluent that detects the sink of the satisfaction signal based on the provided state, s at time, t and returns a truth value.

Definition 2. The degree of an agent's autonomy during the action selection stage evolves by the combination of the degree of both its knowledge and authority to respond to the event via selecting proper action(s).

$$\forall\, e_x \,\exists\, ag_i\text{: } know\,(e_x) \wedge can\,(e_x) \Rightarrow a_i = \left(\frac{k_{e_x}+c_{e_x}}{2}\right) * \gamma \Leftrightarrow min[k_{e_x}, c_{e_x}] \geq T \quad (6)$$

where the function *min* selects the minimum between k_{e_x} and c_{e_x}, and compares the result with the autonomy threshold, T, at the point I (see Figure 4).

Definition 3. The degree of autonomy of an agent during autonomy layer selection and action processing stages in an event is the degree of its authority to handle the event via implementing the selected action(s) in an appropriate autonomy layer.

$$\forall\, e_x \,\exists\, ag_i\text{: } can\,(e_x) \Rightarrow a_i = c_{e_x} * \gamma \quad (7)$$

Definition 4. An agent autonomy layer selection choice is the ratio of the agent autonomy level to the required layer autonomy level:

$$ag_i \equiv L_l \Leftrightarrow \frac{a_i}{a_l} \geq 1 \quad (8)$$

where a_l is layer L_l autonomy level and ≡ denotes the ability of ag_i to use L_l.

We approached that, it is fundamentally necessary to identify the variable level of autonomy of an agent during its interactions in an environment. At the observation and reasoning of an event stages, the agent with the higher knowledge to deliberate on what is observed is more qualified than the agent with lesser knowledge. In the action selection stage, both the knowledge and authority attributes are involved to determine which agent to act as different agent might select different actions to perform the same task. During the execution of the action, the agent autonomy condition is built upon its authority to act such as accessing specific resources to perform the action. If at any time, there are no qualified agents in the system that can make a decision, the system requests for third party intervention.

4 Case Study

To closely look at the aspects of agent autonomy, we need to place the agent in an environment where events occur, and objects and other agents are present. To illustrate our conception, we give an example of an agent-based drone system. This case study is intended to show the effectiveness of the knowledge and authority as criteria of autonomy measurement. Here, we make no assumptions related to the agent's internal state, and coordination and cooperation among agents in the system.

Let's say that during a mission, when the drone observes an obstacle, say a building X (object) in its path. The recognition of an obstacle X is an event, say e_x; then the corresponding task is to avoid crashing into X, i.e. the desire D of the system. The D can be configured if and only if a drone agent ag_1, is able to perceive object X, which means, it has knowledge of what X is. The possible actions to achieve the D are changing the drone's direction to the right R_{ac}, left L_{ac}, up U_{ac} or down D_{ac}, which forms the four options of the drone's action. Therefore, the decision of the ag_1 is to select one of the actions $< R_{ac}, L_{ac}, U_{ac}, D_{ac} >$ to be executed. In this case, the system needs to treat the situation of changing the drone direction differently, as crashing into

X is not a part of the plan. Therefore, the action of "changing the direction" is no longer deemed as a basic task involving low-level actions.

If the drone agent ag_1 recognizes the building is a house or hotel, H, then it will drop the action D_{ac} from its options during the reasoning process as going down to avoid the H will cause the drone to crash into the ground. Contrarily, if the building is a bridge, B, then the ag_1 must drop the actions R_{ac} and the L_{ac} from its options because, it cannot avoid the crashing into X. If the drone agent ag_1 does not know what a bridge is, and treats a bridge like a hotel, then instead of dropping action R_{ac} and L_{ac}, it will drop the action D_{ac} from its options. Therefore, the drone agent ag_1 cannot be allowed to decide because its lack of knowledge to distinguish between a bridge and a hotel. Let's say there is another drone agent, ag_2, that can differentiate between H and B because of it has better knowledge about the objects in the environment. As a result, drone agent ag_2 will opt for the correct action. This leads ag_2 to be authorized to direct drone ag_1 on the direction to steer its course.

What if there is another drone agent, ag_3, that simultaneously acts with ag_2 to handle event e_x, and selects the task to destroy X as it is equipped with fire power, and the corresponding action is to fire a missile at X. Essentially, this type of actions requires high-level of authority which absolutely leads the system to block such behaviour if the ag_3 does not have the authority to do so. Hence, we recognize that knowledge and authority as the key factors that determine autonomy distribution.

5 Conclusion and Future Work

The principle of involving humans and software agents to carry out some system's initiative manifests the notion of adjustable autonomy. However, building a mechanism that controls the behaviour of agents toward an optimized adjustment is the main challenge with the modelling of adjustable autonomy, especially, in dynamic systems. One promising solution is to establish an autonomy measurement mechanism that assists the system in the autonomy distribution process.

In this paper, we propose one such mechanism that can consistently distribute the autonomy to an agent with the best qualification to perform a task. Our aim is to enable the control to be subjected to external and internal influences to eliminate any deficiencies and conflicts of agents' authority that might accrue and cause dramatic situation. We show the importance of separating the autonomy to dimensions of different levels to confined autonomy adjustment in a specific band. Consequently, we demonstrate the possible relationship between the knowledge and authority and their effect on autonomy distribution. We further show the possibility of an ultimate situation of autonomy where system autonomy management is optimized.

In our future work, we shall focus on investigating a situation awareness mechanism that is compatible with the proposed Layered Adjustable Autonomy (LAA) model specifications. The proposed concepts of the LAA are to be applied to implement an agentised drone system for refinement and validation purposes.

References

1. Ball, M., Callaghan, V.: Managing Control, Convenience and Autonomy: A Study of Agent Autonomy in Intelligent Environments. In: Special Issue on Agent-Based Approaches to Ambient Intelligence. AISE series, pp. 159–196. IOS Press (2012)
2. Bradshaw, J.M., Feltovich, P.J., Jung, H., Kulkarni, S., Taysom, W., Uszok, A.: Dimensions of Adjustable Autonomy and Mixed-Initiative Interaction. In: Nickles, M., Rovatsos, M., Weiss, G. (eds.) AUTONOMY 2003. LNCS (LNAI), vol. 2969, pp. 17–39. Springer, Heidelberg (2004)
3. Sellner, B., Heger, F., Hiatt, L., Simmons, R., Singh, S.: Coordinated Multi-agent Teams and Sliding Autonomy for Large-scale Assembly. Proceeding of the IEEE 94(7), 1425–1444 (2006)
4. Brookshire, J., Singh, S., Simmons, R.: Preliminary Results in Sliding Autonomy for Coordinated Teams. In: Proceedings of the AAAI 2004 Spring Symposium, CA (2004)
5. Roehr, T.M., Shi, Y.: Using a Self-confidence Measure for a System-initiated Switch between Autonomy Modes. In: Proceedings of the 10th International Symposium on Artificial Intelligence, Robotics and Automation in Space, Sapporo, Japan, pp. 507–514 (2010)
6. Sheridan, T.B., Verplank, W.L.: Human and Computer Control of Undersea Teleoperators. Technical report, MIT Man-Machine Systems Laboratory, Cambridge, MA (1978)
7. Mercier, S., Tessier, C., Dehais, F.: Adaptive Autonomy for a Human-Robot Architecture. In: 3rd National Conference on Control Architectures of Robots, Bourges (2008)
8. Mostafa, S.A., Ahmad, M.S., Annamalai, M., Ahmad, A., Gunasekaran, S.S.: A Dynamically Adjustable Autonomic Agent Framework. In: Rocha, Á., Correia, A.M., Wilson, T., Stroetmann, K.A. (eds.) Advances in Information Systems and Technologies. AISC, vol. 206, pp. 631–642. Springer, Heidelberg (2013)
9. Myers, K.L., Morley, D.N.: Human Directability of Agents. In: K-CAP 2001. ACM 1-58113–380 (2001)
10. Brainov, S., Hexmoor, H.: Quantifying Relative Autonomy in Multiagent Interaction. In: Hexmoor, H., Castelfranchi, C., Falcone, R. (eds.) Agent Autonomy, pp. 55–74. Kluwer, Dordrecht (2002)
11. Brainov, S., Hexmoor, H.: Quantifying Relative Autonomy in Multiagent Interaction. In: Proc. of the IJCAI 2001 Workshop on Autonomy, Delegation, and Control: Interacting with Autonomous Agents, Seattle, WA, pp. 27–35 (2001)
12. Mostafa, S.A., Ahmad, M.S., Annamalai, M., Ahmad, A., Gunasekaran, S.S.: A Conceptual Model of Layered Adjustable Autonomy. In: Rocha, Á., Correia, A.M., Wilson, T., Stroetmann, K.A. (eds.) Advances in Information Systems and Technologies. AISC, vol. 206, pp. 619–630. Springer, Heidelberg (2013)
13. Mostafa, S.A., Ahmad, M.S., Annamalai, M., Ahmad, A., Basheer, G.S.: A Layered Adjustable Autonomy Approach for Dynamic Autonomy Distribution. In: The 7th International KES Conference on Agents and Multi-agent Systems Technologies and Applications (KES-AMSTA), vol. 252, pp. 335–345. IOS Press, Hue City (2013)
14. Parasuraman, R., Sheridan, T., Wickens, C.: A Model for Types and Levels of Human Interaction with Automation. IEEE Transactions on Systems, Man and Cybernetics, Part A 30(3), 286–297 (2000)
15. Barber, K.S., Martin, C.E.: Agent Autonomy: Specification, Measurement, and Dynamic Adjustment. In: Proceedings of the Autonomy Control Software Workshop at Autonomous Agents (Agents 1999), Seattle, WA, pp. 8–15 (1999)

16. Mercier, S., Dehais, F., Lesire, C., Tessier, C.: Resources as Basic Concepts for Authority Sharing. In: Humans Operating Unmanned Systems, HUMOUS 2008 (2008)
17. Huber, M.J.: Agent Autonomy: Social Integrity and Social Independence. In: Proceedings of the International Conference on Information Technology ITNG 2007, pp. 282–290. IEEE Computer Society, Los Alamitos (2007)
18. Mark, E., Anderson, J., Crysdale, G.: Achieving Flexible Autonomy in Multi-Agent Systems using Constraints. Applied Artificial Intelligence: An International Journal 6, 103–126 (1992)
19. Anderson, J., Evans, M.: Supporting Flexible Autonomy in a Simulation Environment for Intelligent Agent Designs. In: The Proceedings of the Fourth Annual Conference on AI, Simulation, and Planning in High Autonomy Systems, pp. 60–66. IEEE, Tucson (1993)
20. Luck, M., D'Inverno, M., Munroe, S.: Autonomy: Variable and Generative. In: Hexmoor, H., Castelfranchi, C., Falcone, R. (eds.) Agent Autonomy, pp. 9–22. Kluwer (2003)
21. Castelfranchi, C.: Guarantees for Autonomy in Cognitive Agent Architecture. In: Wooldridge, M.J., Jennings, N.R. (eds.) ECAI 1994 and ATAL 1994. LNCS (LNAI), vol. 890, pp. 56–70. Springer, Heidelberg (1995)
22. Moore, R.C.: A Formal Theory of Knowledge and Action. In: Allen, J.F., Hendler, J., Tate, A. (eds.) Readings in Planning, pp. 480–519. Morgan Kaufmann (1990)
23. Laird, J.E., Rosenbloom, P.S.: The Evolution of the Soar Cognitive Architecture. In: Steier, D.M., Mitchell, T.M. (eds.) Mind Matters: A Tributr to Allen Newell, pp. 1–50 (1996)

Emergence of Norms in Multi-agent Societies: An Ultimatum Game Case Study

Mihai Trăşcău[1], Marius-Tudor Benea[1,2], Teodor Andrei Tărtăreanu[3], and Şerban Radu[1]

[1] Computer Science Department, University Politehnica of Bucharest, Romania
[2] LIP6, University Pierre and Marie Curie, France
[3] Axway (NYSE Euronext: AXW.PA)
{mihai.trascau,serban.radu}@cs.pub.ro, marius-tudor.benea@lip6.fr, andrei.tartareanu@gmail.com

Abstract. This article studies the problem of norm emergence in the context of multi-agent systems (MAS). The agents of a certain MAS repeatedly play the Ultimatum game, by means of individual social interactions. The simulations are run and, afterwards, studied, for different learning strategies (a normative advisor model and a social model), and for different structures of the society (some of the most popular network topologies are used).

Keywords: multi-agent systems, norm emergence, social interactions, learning strategies, network topologies.

1 Introduction

Norms have a very important role in any society. They tend to be generally accepted by all the members of the society, as they are the result of a complex process of emergence that starts with the simplest social interactions between the individuals and continues as the individuals repeatedly interact in the framework of a given scenario. Due to this bottom-up process of evolution, the norms fit better the needs of the population, as opposed to the top-down mechanism of enforcement of the laws. Moreover, fewer resources need to be allocated by a central authority for their creation and assimilation by the society as in the case of the laws. And this owes to the collective efforts made by the whole society for the same purpose. However, there is still a cost, the lack of precision. For this reason, it becomes vital to understand the mechanisms behind the emergence of norms.

There are many aspects of agent interactions which influence norm emergence, like the underlying network structure, the representation of norms and the propagation mechanisms, the types of social interactions and their frequency, or the normative strategies of the agents. In this article we intend to study two of these aspects for the problem of norm emergence, topologies (which are a very important factor in the evolution of a society and has a strong impact on the emergence of norms; moreover, there isn't too much research on the topic in

A. Bădică et al. (eds.), *Recent Developments in Computational Collective Intelligence*, Studies in Computational Intelligence 513,
DOI: 10.1007/978-3-319-01787-7_4, © Springer International Publishing Switzerland 2014

the context of norm emergence) and normative strategies (we believed they had a very important role, too, in the process of norm emergence and, afterwards, the results proved it), and to present some experimental results in support of our observations. These results are obtained after multiple simulations using a scenario based on Ultimatum Game.

The structure of the paper is the following one: First, we present some related works, in Section 2. Further, in Section 3 we describe the mechanisms used for norm emergence and propagation and the model of the agents. In Section 4 we make an analysis of the topologies used. An experimental results analysis is presented in Section 5. Finally, in Section 6 conclusions are drawn and ideas for extending the current study are presented.

2 Related Work

Research in normative multi-agent systems has mainly two directions. The first direction focuses on normative system architectures, norm representation, norm adherence and the associated punitive or incentive measures. Lopez [1] et al. proposed and designed an architecture for BDI agents, while Boella [2] et al. described a distributed architecture for normative agents.

The second direction of research (also followed by this work) is related to the emergence of norms. With their decentralized behavior-shaping effect, the emergence of norms becomes a potentially scalable, adaptive control mechanism. While works belonging to this direction, like [3–6], study this phenomenon, of norm emergence, too, they do it from different perspectives.

None of the works mentioned above study in depth the influence of the topologies of the networks of interactions between the agents on emergence of norms. Mukherjee et al., in [5, 4] treat this problem only a little bit, through what they call "spatially constrained interactions", taking the form of agent neighborhoods. Pujol's PhD thesis, [7] is another example, but we chose to further extend his study with some other topologies, *complete graphs* and *random regular graphs.* While the complete graphs are a special case of regular graphs worth considering, for the high connectivity between the nodes, the choice for random regular graphs (both random and regular) considered the fact that the topologies most resembling real world societies, scale free and small-world, have a certain degree of randomness and its influence should be thoroughly understood.

Moreover, all of the works specified in the second paragraph use simple scenarios for the experiments, in which the decisions made by the agents are based on simple (2x2) payoff matrices. We consider that the Ultimatum game, in which the agents have to learn some more complex strategies in order to play a difficult negotiation game, allows us to study a scenario that is closer to a larger class of real world problems. Last, but not least, we introduced an ingredient often present in the real world's social interactions, the role models. We study, in our simulations, important aspects as their role and their power of influence.

Another work studying the process of norm emergence in multi-agent systems, given the ultimatum game, is Savarimuthu's [8]. We tried to obtain some extra results, first by testing the system we implemented on various network topologies, then by changing how the agents decided on their role-model. In our work they choose the role-model from their topological neighbors, based on the success rate of every neighbor after each epoch.

3 Models and Mechanisms

In this section we present, in depth, the scenario used, the agent models and the architecture of the simulation.

3.1 Simulation Scenario

Our choice is focused on the "Ultimatum game", a well known game often used in economic experiments, which Slembeck [9] considers to be probably the most frequent studied bargaining game. This is based on an one-off interaction between two players that play for a fixed sum of money (or points, credits, etc.). The first player proposes a distribution of the sum for both players, while the second player can choose to either accept or reject the offer, in the latter case both players receiving nothing.

3.2 Agent Models

The multi-agent systems used in the simulation are based on two normative strategy mechanisms: a *normative advisor* approach and a *social* approach. These two mechanisms are studied separately, in order to highlight the differences between them and, by comparison, the importance of the social approach and of the role model ingredient (see Section 2).

For the approach based on the normative advisor, there are two types of agents: the *players* and the *advisor*. In a given simulation there are many players and a single advisor. The latter does not participate in the game, but plays the role of an observer and gives advices to the other agents. In the social approach each agent in the society chooses its own role model, based on an overview of the performances of its neighbors.

Each playing agent can have a proposer role in which he is able to propose a norm to other agents in the society, or a decider/acceptance role to accept or not a norm that was proposed to him. Each agent can take one role or another.

A norm is composed of two sub-norms, one used in the proposal and the other in the acceptance role. Each sub-norm provides a *min* and a *max* value, which are used by the agents in order to formulate proposals or to decide wether to accept a proposal or not, according to their role for a given interaction. A certain degree of autonomy is simulated using probabilities (later described in this section).

The Normative Advisor Approach. In order to simulate various levels of autonomy among agents, two norms have been used: a *personal* norm and a *collective* (that applies to the whole society) norm which is received as an advice from the normative advisor. Agents are free to accept or reject the advised norm and if, based on a probability, they accept it, the personal norm will still be the one used, but after it will be modified with respect to the collective norm, according to a formula presented later.

After each turn, the ***playing agent*** receives a new norm advice from the normative advisor (the collective norm), which can be considered or not for modifying the personal norm w.r.t. a probability $prob_{autonomy} \in [0, 1]$ that describes the degree of autonomy of the agent. If the agent accepts the advice, its personal norm is updated as follows:

$$\begin{cases} min_P = min_P + \alpha \cdot (min_C - min_P) \\ max_P = max_P + \alpha \cdot (max_C - max_P) \end{cases}$$

where $\alpha \in [0, 1]$ is the acceptance factor, min_P, max_P are the interval limits of the personal norm and min_C, max_C are the interval limits of the advised, collective, norm. The $prob_{autonomy}$ probability is specific to each agent and its value is set before the beginning of the simulation, randomly. We mention that these formulas are used for both the proposal and the acceptance sub-norms, according to the role played by the agent.

When playing the proposer role, the agent randomly chooses a value from the interval described by the norm used, based on a uniform distribution. On the other hand, when playing the decider role, the agent decides to accept the offer only if the proposed values lie within its acceptance interval.

The ***normative advisor*** is the one that provides coherence and coordination through its advices that are based on analyzing the results of the interactions. It collects information from each agent when the epoch ends, which contains the entire round history. The history contains information about each interaction (e.g., proposed sum, received offer, outcome). Based on these values the normative advisor computes the ratio between the number of successes and the total number of times played for each value. These values are inherently prone to solitary peaks, especially when considering values played only a few times but with high success rates. In order to avoid these peaks, the normative advisor has been designed to search for the longest interval of the amounts of money with the maximum sum of their successes. As the values can be both positive and negative, the mean value of the success rates is subtracted from each success rate of each value, allowing the selection of the longest interval (described above) to be applied on the resulting values.

After the maximum sum subsequence is determined the normative advisor, which keeps the collective norm, modifies its values accordingly:

$$\begin{cases} min_P = min_P + \beta \cdot (min_{new} - min_P) \\ max_P = max_P + \beta \cdot (max_{new} - max_P) \end{cases}$$

with $\beta \in [0, 1]$ being the modifier factor, min_P, max_P being the old interval limits and min_{new}, max_{new} being the newly computed interval limits. These new values of the collective norm are then distributed to each playing agent, at the beginning of the next round.

The Social Approach. In the social approach, the ***playing agent*** (also the single agent type) adopts a normative strategy which takes into consideration the most successful neighbor. After each epoch, inquiries are sent to each of the neighbors, for which replies containing the success rate and the personal norm of the inquired agent are received. Thus, the agent selects the neighbor with the highest success rate and adopts the advised norm using the exact mechanism as for the collective norm received from the advisor, in the normative advisor approach.

When computing how successful a playing agent is, we have designed the social model considering, separately, the number of successful game interactions from the total number of interactions per epoch, in either role played, and the actual profit the agent realizes each epoch. The profit is computed by summing up the scores the agent receives in successful interactions, in either role played.

In every other aspect, this model is designed, for playing agents, as in the case of the normative advisor model. This enables simulations to reveal normative strategy differences concerning norm emergence speed, norm types which emerge and general wellness of the society, as will be described in Section 5.

3.3 Simulation Architecture

As this research is focused on studying norm emergence with respect to topologies and norm strategy, the need for an easy to control and analyze architecture arises. This architectural need leads to the decision of introducing a centralized entity, called Facilitator. It is assigned to choose pairs of two agents for interactions (according to the given topology), before each epoch, synchronize the game steps and epochs and collect and store data regarding norm values from the agents, after each round. It doesn't have any influence on the way norms emerge. It is only part of the simulation platform and facilitates the execution.

Each epoch, the agents interact at least n times (where n is a positive integer), thus giving the chance for all agents to interact. The interactions are simple, having the following form: the proposer generates a proposal and sends it to the decider. The decider sends an answer informing the proposer of its decision to accept or reject the proposal. Both agents update their history according to the messages received.

4 Network Topologies

Although our main goal was to determine the role of social interactions in the process of norm emergence, we found that a good approach would be to consider the influence of topologies too, as they influence the norm propagation. In this section we describe the network topologies used as a basis for the simulations we ran. All graphs described here are undirected.

Regular Graphs are graphs of the form $C_{N,k}$, where N is the number of nodes and k is the connectivity. Considering that the nodes are in a vector of N elements, each node i, is adjacent to nodes $(i-j)\%N$ and $(i+j)\%N$, where $1 \leq j \leq k$. Regular graphs are highly clustered just like any complex networks. In our experiments we used strongly regular graphs.

Complete Graphs are a particular case of regular graphs, in which a node is adjacent with all the other nodes of the network.

Random Graphs (Erdos-Rnyi model), [10], are built starting from N nodes, then uniformly selecting pairs of nodes and adding edges between them. This process repeats m times. Random graphs usually have short path lengths, but, unlike regular graphs, they are not highly clustered.

Random Regular Graphs resemble regular graphs (the nodes are linked randomly) with the restriction that all nodes have the same given degree, d.

Small-World Graphs (Watts-Strogatz model), [11], lie between strongly regular graphs and random graphs, depending on a probability p. Starting from a strongly regular graph, in k steps, where k is the connectivity, for each node i of the graph, the edge between it and the node $(i+j)\%N$, where $1 \leq j \leq k$, is rewired with probability p. There is a $1-p$ probability for the link to be left intact. This way, if we set the probability p to 0 we obtain a strongly regular graph, and if we set p to 1 we generate a random graph.

The graph generated when $0 \leq p \leq 1$ is named "small-world", with analogy to the small-world phenomenon, described by Stanley Milgram, [12]. Its main characteristics are that it is highly clustered, like the strongly regular graphs, and it presents short path lengths, like the random graphs.

Scale Free Graphs, [13], introduced two extra "ingredients", as compared to the complex networks available before them: growth and preferential attachment. Growth refers to the process of starting from a small set of initial nodes m_0 and then growing by adding a new node in each step, which is then connected to m nodes, using a preferential probability. This preferential probability is of the form $\prod(k_i) = \frac{k_i}{\sum_j k_j}$, where k_i is the degree of the node i.

Close similarities were discovered by Barabasi et al. between the structure of many real-world societies and scale free networks. So, in a further study, [14], Albert and Barabsi studied many real world topologies, and managed how to construct scale-free graphs with similar structures.

Scale free networks are dominated by a few number of hubs, but they do not present the small world property. Also, scale free networks follow a power law distribution. Thus, the probability that a node interacts with k other vertices is $P(k) \sim k^{-\gamma}$. In the extended model [15], Barabsi et. al. discovered a way to make γ vary (in the first model it's value was 3), by introducing two more probabilities: p for adding m links between old nodes, and q for rewiring m old links. The remaining $1-p-q$ is the probability of adding a new node, just as in the previous model. In the extended model, the preferential probability is $\prod(k_i) = \frac{k_i+1}{\sum_j k_j+1}$. When p=q, $P(k) \propto [k+1]^{-\frac{2m(1-q)+1-2q}{m}+1}$. We can easily find $p=q$ for a given γ so, this way we created, for our experiments, networks of given kinds, as described in [14] (e.g., $\gamma_{citation} = 3.0 \pm 0.1, \gamma_{www} = 2.1 \pm 0.1$).

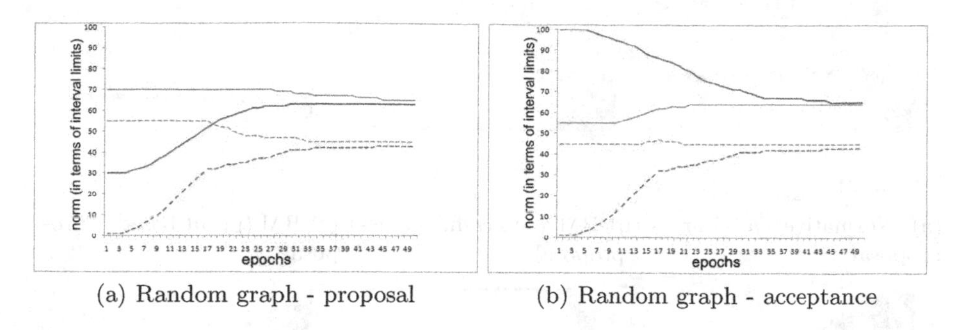

(a) Random graph - proposal (b) Random graph - acceptance

Fig. 1. Emerging norms. blue=benevolent, red=selfish, continuous=maximum, dashed=minimum.

Table 1. Parameters used to generate the networks

Network type	Parameters
Complete	N = 100
Random	N = 100, m = 1000
Strongly regular	N = 100, k = 10
Random regular	N = 100, d = 20
Small-world	N = 100, k = 10, p = 0.5
Barabasi citation, $\gamma = 3$	N = 100, $m_0 = 10$, m = 10, p = q = 0.0455
Barabasi www, $\gamma = 2.1$	N = 100, $m_0 = 10$, m = 10, p = q = 0.4545
Barabasi math, $\gamma = 2.5$	N = 100, $m_0 = 10$, m = 10, p = q = 0.2727

5 Simulation Results

In order to test the norm mechanisms the simulation scenario was devised using a society comprised of 50 "selfish" agents and 50 "benevolent" agents. This society structure was tested with all the topologies presented in Section 4, generated w.r.t. their properties, using the parameters shown in Table 1. The last three network types are Barabsi scale-free networks similar to some real world networks, as studied in [14]. The simulation ran for 100 turns, enough time for the society to reach norm emergence, for which the threshold was considered to be 99%.

The first series of tests conducted to similar results regardless of the topology used. For this reason Figure 1 illustrates only one topology, the other ones looking almost identical. This can be due to the small number of nodes in the network, or due to the high value of the mean degree of the nodes (which is about 20 except for the complete graph), related to the total number of nodes. In Figure 1 (as well as in Figure 2), the horizontal axis represents the number of epochs, while the vertical axis represents norms in terms of intervals which reflect the state of the entire society.

Employing the two different playing agent models, one used in conjunction with the normative advisor and the other based on the local neighborhood of

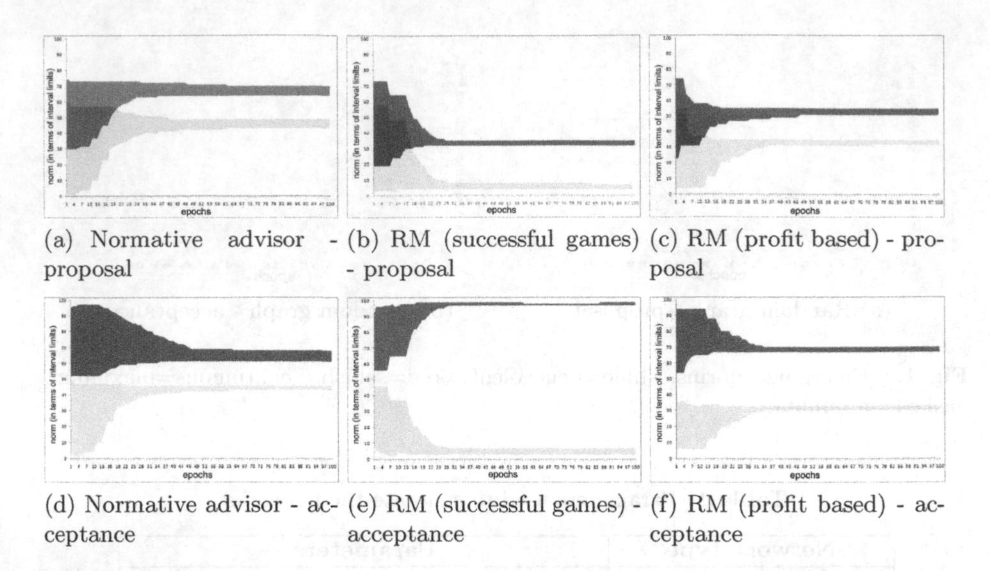

(a) Normative advisor - proposal (b) RM (successful games) - proposal (c) RM (profit based) - proposal

(d) Normative advisor - acceptance (e) RM (successful games) - acceptance (f) RM (profit based) - acceptance

Fig. 2. Subnorm intervals. RM=Role model, black=maximum threshold, grey = minimum threshold.

the agents, led to different results in the game. Although the broad view over the state of the society when using the normative advisor model, together with the advisor itself, helps establishing more compact intervals for the norms, the norm emergence is reached more slowly than in the social approach. This can be explained by the fact that in the second case each agent adapts to its immediate neighborhood, not to the general trend of the society. This rapidly creates very successful agents which, being repeatedly elected as role-models, help spread the norm at a faster pace. The larger intervals observed in the social approach can be interpreted as a result of the limited view each agent has over the society.

We were also able to conclude that, considering the two sub-norms defined, the acceptance norm interval adjustment is more drastic within the same number of epochs as opposed to the proposal norm interval, mostly because "benevolent" agents tended to constantly offer higher deals which were accepted by both fair and unfair agents, and "self-interested" agents tended to raise their lower acceptance threshold faster as they tried to "speculate" on other agents.

In Figure 2, proposal and acceptance sub-norm interval evolution over the epochs, for all agents in the society, are presented either when using normative advisors or in the social approach. Also, for the social model, results for both success rate computing mechanisms are presented. The two approaches of the social model yield quite different results, the first one (successful games) favors norms considered to be "unfair" to emerge, while the latter (profit based) tips the balance towards the emergence of "benevolent" norms. Furthermore, data gathered from the agents in the two approaches reveal differences in the total wellness of the society, as is described in Figure 3. The social approach leads to an increase of the wellness in the society by 20%.

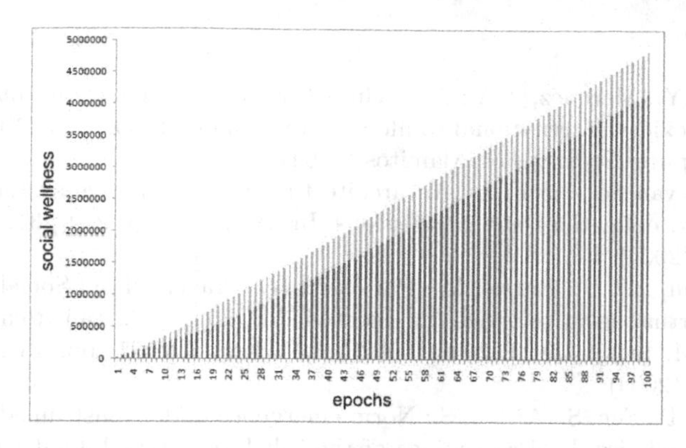

Fig. 3. Wellness gain for the social approach

6 Conclusion and Future Work

In this paper we described and presented the results, together with an interpretation of them, for a series of multi-agent system simulations, in which the agents repeatedly played the Ultimatum game at individual scale with other members of the society. The simulations were made in the context of a normative agent model and of a social learning model and were based on different underlying topologies of interactions. With respect to the two types of normative strategies, simulation results indicate a society wellness increase in the case of the social learning mechanism of up to 20% as opposed to the normative advisor mechanism case.

For the tests run there weren't significant differences between the topologies, but we expect to obtain different results for networks generated using different parameters. As future work, considering this aspect, we plan to see if these results persist for topologies generated with different parameters. We expect to draw some interesting conclusions by generating topologies with lower mean degrees for their nodes. We also wish to modify the simulation platform so that simulations with considerably larger MAS could be run. We are interested in studying the influence of topologies on these larger societies.

Other future work plans include designing and implementing a framework for developing social interaction games. This framework is intended to facilitate network creation, testing and analysis, allowing for a faster deployment of simulation scenarios.

Acknowledgement. This work was funded by the project ERRIC (Empowering Romanian Research on Intelligent Information Technologies), number 264207/ FP7-REGPOT-2010-1.

References

1. Lopez, F.L.Y., Marquez, A.A.: An architecture for autonomous normative agents. In: Fifth Mexican International Conference in Computer Science (ENC), p. 96173. IEEE Computer Society, Los Alamitos (2004)
2. Boella, G., van der Torre, L.: An architecture of a normative system: counts-as conditionals, obligations and permissions. In: AAMAS, pp. 229–231. ACM Press, New York (2006)
3. Savarimuthu, B.T.R., Purvis, M., Purvis, M.K., Cranefield, S.: Social norm emergence in virtual agent societies. In: Baldoni, M., Son, T.C., van Riemsdijk, M.B., Winikoff, M. (eds.) DALT 2008. LNCS (LNAI), vol. 5397, pp. 18–28. Springer, Heidelberg (2009)
4. Mukherjee, P., Sen, S., Airiau, S.: Norm emergence under constrained interactions in diverse societies. In: Proceedings of the 7th International Joint Conference on Autonomous Agents and Multiagent Systems, AAMAS 2008, vol. 2, pp. 779–786. International Foundation for Autonomous Agents and Multiagent Systems, Richland (2008)
5. Mukherjee, P., Sen, S., Airiau, S.: Norm emergence in spatially constrained interactions. In: Working Notes of the Adaptive and Learning Agents Workshop at AAMAS, vol. 7 (2007)
6. Sen, S., Airiau, S.: Emergence of norms through social learning. In: IJCAI 2007: Proceedings of the 20th International Joint Conference on Artifical Intelligence, pp. 1507–1512 (2007)
7. Pujol, J.M.: Structure in Artificial Societes. PhD thesis, Universitat Politecnica de Catalunya, Departament de Llenguatges i Sistemes Informatics, PhD Program: Artificial Intelligence, Barcelona (2006)
8. Savarimuthu, B.T.R., Purvis, M., Cranefield, S., Purvis, M.: How do norms emerge in multi-agent societies? mechanisms design. The Information Science Discussion Paper Series, Number 2007/01 (February 2007) ISSN 1177-455X
9. Slembeck, T.: Reputations and fairness in bargaining - experimental evidence from a repeated ultimatum game with fixed opponents. Experimental 9905002, EconWPA (1999)
10. Erdos, P., Rényi, A.: On the evolution of random graphs. Publ. Math. Inst. Hungar. Acad. Sci. 5, 17–61 (1960)
11. Watts, D.J., Strogatz, S.H.: Collective dynamics of 'small-world' networks. Nature 393 (June 4, 1998)
12. Travers, J., Milgram, S.: An Experimental Study of the Small World Problem. Sociometry 32(4), 425–443 (1969)
13. Barabasi, A.L., Reka, A.: Emergence of scaling in random networks. Science 286 (October 15, 1999)
14. Réka, A., Barabási, A.L.: Statistical mecanics of complex networks. Reviews of Modern Physics 74 (January 2002)
15. Barabási, A.L., Réka, A.: Topology of evolving netwotks: Local events and universality. Physical Review Letters 85(24) (December 11, 2000)

A Fractal Based Generalized Pedagogical Agent Model

Soufiane Boulehouache[1,2], Remdane Maamri[1], Zaidi Sahnoun[1], and Alla Larguet[2]

[1] LIRE Laboratory, Constantine 2 University, Algeria
[2] Université 20 août 1955-Skikda, Algeria
sboulehouache@yahoo.com, sahnounz@yahoo.fr, rmaamri@umc.edu.dz

Abstract. *Pedagogical Agent* (*PA*) is a tendency in human learning systems (*HLS*). However, the used Agent models are unable to decrease the complexity/performance ratio introduced by the modeling and the manipulation of multi-knowledge domains (Domain Model, Student Model, Pedagogical Model, etc). Thus, within this paper, we propose a *Component based Pedagogical Agent Model* that we claim makes effective and flexible the *Pedagogical Agents* building and maintaining. The *PA* building is simplified to an assembling of a set of pre-built sub-components that is directed by the *PA ADL Fractal Abstract Description*. Also, maintaining a sub set of sub-components means disconnecting them, removing them and replacing them by other versions those integrate the new required functionalities and/or tuning. Therefore, the building and the maintaining efficiency and flexibility are achieved by the separation of the abstract description of the *PA* from its fulfilling.

Keywords: Pedagogical Agent, Component-Based Agents, the Fractal model, and the ADL Fractal.

1 Introduction

The Pedagogical Agent[1] *is* a new trend in human learning systems. It is defined as an educational system that has learning strategies and that is formed by *Intelligent Agents* [19]. *Vladan Devedzic and al.* [7] state that over the last several years, there has been significant interest within the Intelligent Tutoring System (ITS) research community for applying *Intelligent Agents* in design and deployment of ITSs. This was lead to interesting results have been achieved by pedagogical agents [25]. However, the usual Agent models and building tools were insufficiently decreasing the complexity of the *PAs* building and maintaining. Because, the usual Agent models keep a heavy coupling between the *PAs's* modules that is inappropriate to efficiently building and maintaining their structural sub-parts. So, the challenge initiated by *Carine Webber et al.* [18] in looking for a generic model of teaching and/or learning and referring to a static and academic view of the knowledge to be taught still open.

[1] Examples of PA are: ADELE (*Agent for Distance Learning Environment),* STEVE (**S**oar Training **E**xpert for **V**irtual **E**nvironments), etc.

A. Bădică et al. (eds.), *Recent Developments in Computational Collective Intelligence*,
Studies in Computational Intelligence 513,
DOI: 10.1007/978-3-319-01787-7_5, © Springer International Publishing Switzerland 2014

In fact, the *Component* seems suitable to decrease the complexity of the *Pedagogical Agent's building and maintaining*. It is a paradigm that aims at constructing systems using a pre-defined set of software components explicitly created for reuse [9]. According to *Fabresse et al.*, the Component is an approach of designing and building for reuse and by reuse of software components [1]. The main idea of software reuse is to use previous software components to create new software programs [9]. Indeed, it takes less time to buy, and then to reuse, a component besides create, encode, test, and document the debugging purposes [5]. A Component is a unit of composition with contractually specified interfaces and context dependencies only. A software component can be deployed independently and is subject to composition by third parties [3]. Components are not used in isolation, but according to a software architecture that determines the interfaces that components may have and the rules governing their composition [9]. Consequently, a number of Component models are proposed -i.e. the *Fractal Component Model* [6]. They define what components are, how they can be constructed, how they can be composed or assembled, and how they can be deployed [5].

Within this paper we propose a *Fractal based Generalized Pedagogical Agent Model* that aims to increase the modularity of the *Pedagogical Agent* and makes effective and flexible its *building and maintaining*. The most interesting point of this model is the inherited separation of the abstract description of the components from their concrete fulfilling. This assures the efficiency and the flexibility to decrease the complexity of the *Pedagogical Agent.* Furthermore, maintaining the *Fractal based Pedagogical Agent* is simplified to the replacing of a sub set of components by other versions those provide the required functionalities and /or quality.

We discuss the proposition in the remainder of this paper as follows. Second, we introduce the generalized pedagogical agent and its patterns. The section 3 focuses on the Fractal model. Next, we present A Fractal based Generalized Pedagogical Agent. Next, we give a short discussion of the benefits of the propose Fractal based Generalized Pedagogical Agent. The last is a summary and conclusion.

2 The Generalized Pedagogical Agent

Intelligent Tutoring Systems (ITSs) are emerged from the integration of the Artificial Intelligence (AI) technology to educational systems. Their goal was to reproduce the behavior of an intelligent (competent) human tutor who can adapt his teaching to the learning rhythm and other characteristics of the learner [16]. In an attempt to increase the efficiency and the flexibility, the ITS research community focus is on *Pedagogical Agents*[2]. According to *Johnson et al* in [12], practical autonomous agents must in general manage complexity. From a software engineering point of view, agents have advantages of modularity and flexibility [13]. They can be autonomous agents that support human learning by interacting with students in the context of interactive learning environments [7]. They monitor the dynamic state of the learning environment, watching for learning opportunities as they arise. They can support both

[2] For further reading refer to [7, 12, 17, 20, and 18].

collaborative and individualized learning, because multiple students and agents can interact in a shared environment. *Pedagogical Agents* can provide numerous instructional interactions with students, promote student motivation and engagement, and effectively support students' cognitive processes [7].

The figure 1 shows the architecture presented by *Vladan Devedzic* and *Andreas Harrer*, in [7], is a summary of the patterns used in designing *Pedagogical Agents*.

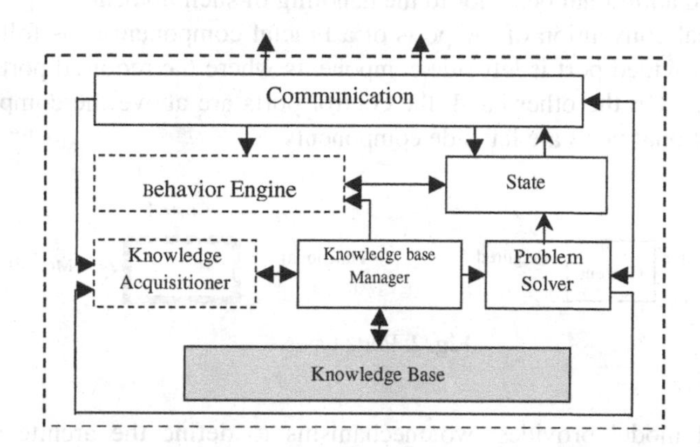

Fig. 1. The Generalized Pedagogical Agent [7]

However, the *Pedagogical Agents* are unable to really decrease the complexity/ performance ratio.

The next section is a brief presentation of the *Fractal Model.* It focuses on the architectural properties necessary for the modeling of the *Generalized Pedagogical Agent.* Thus, the architecture shifts the focus of developers from lines-of-code to coarser grained architectural elements and their overall interconnection structure [31].

3 The Fractal Component Model

Fractal[3] is a component model that its main goals are to implement, deploy and manage complex software systems. A Fractal Component is a runtime entity that is encapsulated, and that has a distinct identity. It can be either primitives or composites. It comprises two parts: a membrane, which provides external interfaces to introspect and reconfigure its internal features (called control interfaces), and a content, which consists in a finite set of other components (called sub-components). The membrane of a component is charged of providing control capacity and can have external and internal interfaces. External interfaces are accessible from outside the component, while internal interfaces are only accessible from the component's sub-components. The membrane of a component is typically composed of several controller and interceptor objects. A controller object implements control interfaces. Typically, a

[3] For further reading refer to [6, 26].

controller object can provide an explicit and causally connected representation of the component's sub-components and superpose a control behavior to the behavior of the component's sub-components, including suspending, checkpointing and resuming activities of these sub-components. Interceptor objects are used to export the external interface of a subcomponent as an external interface of the parent component. They intercept the oncoming and outgoing operation invocations of an exported interface and they can add additional behavior to the handling of such invocations [6, 26].

The graphical convention of the ports of a Fractal component is as following: on one hand, the provided port is left side components where the required port is at right side components. On the other hand, the control ports are above the component box where, the functional ports are left side components.

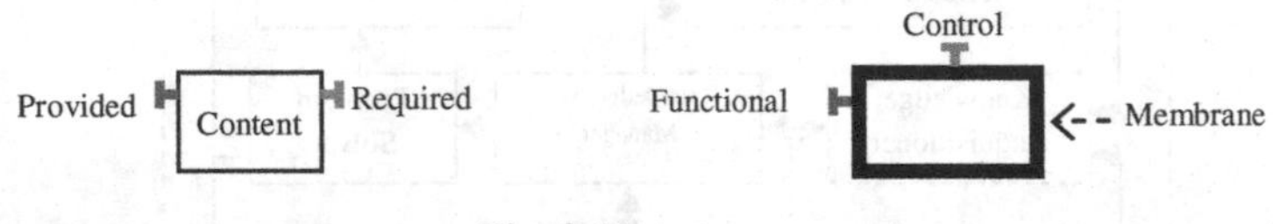

Fig. 2. Ports types

The Fractal model provides two mechanisms to define the architecture of an application: binding between component interfaces, and encapsulation of a group of sub-components in a composite. Fractal supports both primitive bindings and composite bindings. A binding is a normal Fractal component whose role is to mediate communication between other components. A primitive binding is a binding between one client interface and one server interface in the same address space, which means that operation invocations emitted by the client interface should be accepted by the specified server interface. A composite binding is a communication path between an arbitrary numbers of component interfaces [6, 26].

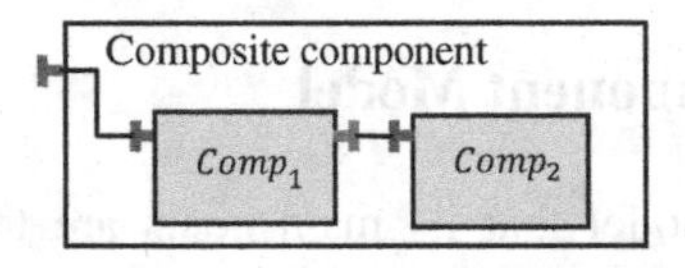

Fig. 3. The Binding of Fractal Components

4 A Fractal Based Generalized Pedagogical Agent Model

The *Fractal Based Generalized Pedagogical Agent* (*FGPA*) is modeled architecturally at a module level. It means that, each module is modeled as a Fractal sub-component of the *GPA Composite Component.* In this sub section we present the different sub-components constituting the *FGPA* with a special emphasize on the goals, interfaces types and their binding.

4.1 The Knowledge Base Component

The *Knowledge Base* is modeled as a composite component that contains the following sub-components: the *Domain Model* (*DM*), the *Student Model* (*SM*) and the *Pedagogical Model* (*PM*). To access and updates the knowledge of the three sub-components, the *Knowledge Base Component* (*KBC*) provides three server interfaces (edmi, esmi and epmi). Also, the three sub-components are exposed to the external address space through three internal client interfaces of the *KBC* using binding. This latter does not have client interfaces because it contains only the necessary knowledge for a *Pedagogical Agent* to work. The *figure 4* shows the *KBC*. Also, in all the next figures, we present the schematic view and the *ADL Fractal* abstract description.

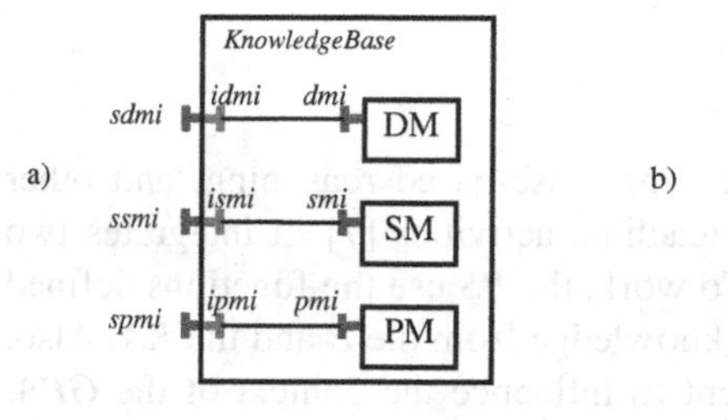

b)

```
< component name="KnowledgeBase"/ >
<interface name="sdmi" role="server" signature="sdmis"/ >
<interface name="ssmi" role="server" signature="ssmis"/ >
<interface name="spmi" role="server" signature="spmis"/ >
<interface name="idmi" role="client" signature="idmis"/ >
<interface name="ismi" role="client" signature="ismis"/ >
<interface name="ipmi" role="client" signature="ipmis"/ >
<component name="dm" definition="DomainModel" />
<component name="sm" definition="StudentModel" />
<component name="pm" definition="PedagogicalModel" />
<binding client="this.idmi" server="dm.dmi" />
<binding client="this.ismi" server="sm.smi" />
<binding client="this.ipmi" server="pm.pmi" />
< /component>
```

Fig. 4. A Knowledge Base Component

4.2 Knowledge Base Manager Component

The *Knowledge Base Manager* (*KBM*) main goal is to control the access and the updates to the *Knowledge Base* [7]. The *KBM* component is constituted of three client interfaces and three server interfaces and it is implemented through content. The server interfaces are used by the *Problem Solver*, the *Behavior Engine* and the *Knowledge Acquisitioner* to access the *Knowledge Base*. However, the three client interfaces are responsible of the access and updates of the three models of the *KBC*.

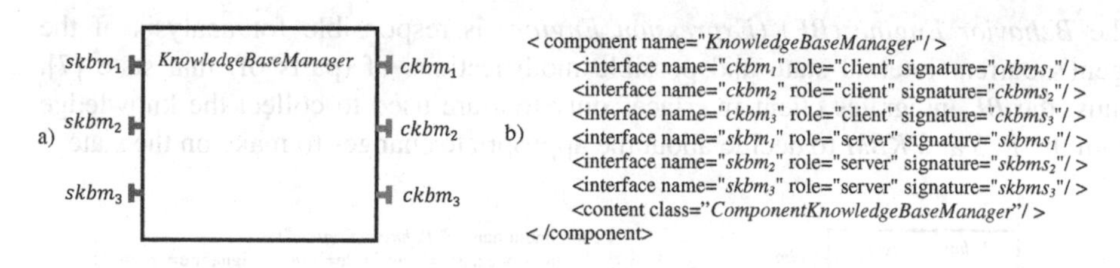

Fig. 5. A Knowledge Base Manger Component

4.3 The Knowledge Aqusitioner Component

The *Knowledge Acquisitioner (KA)* reflects the fact that cognitive pedagogical agents also possess a learning capability that makes possible for the *GPA* to update and modify its knowledge over time, using some machine learning technique(s) [7, 20]. The functionalities provided through the "ska_1" of the *KA* component are used by the *KBM*. The client interfaces are used to bind the *KA* to the KBM, *Behavior Engine (BE)* and the *Perception (P)* components.

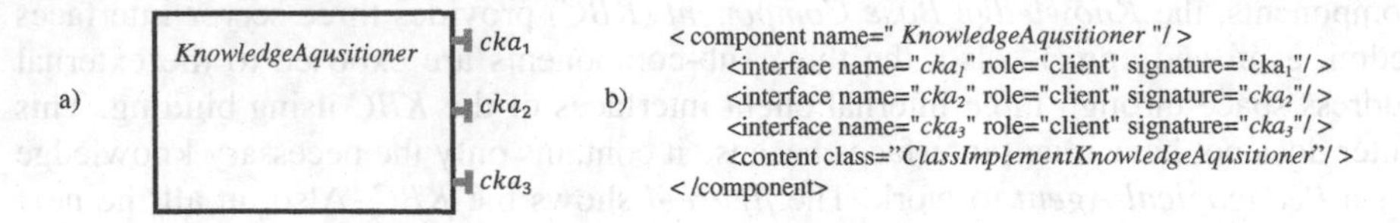

Fig. 6. A Knowledge Aqusitioner Component

4.4 The Problem Solver Component

The Problem Solver (*PS*) can include inferencing, case-based reasoning, and other tasks normally associated with learning and teaching activities [7]. It integrates two server interfaces and three client interfaces. To work, the *PS* use the functions defined in the client interfaces to requests the needed knowledge from the *P* and the *KA*. Also, it uses the binding with the Action component to influence the context of the *GPA*. Furthermore, it provides reasoning capacity to the *KA* and *State (S)* components.

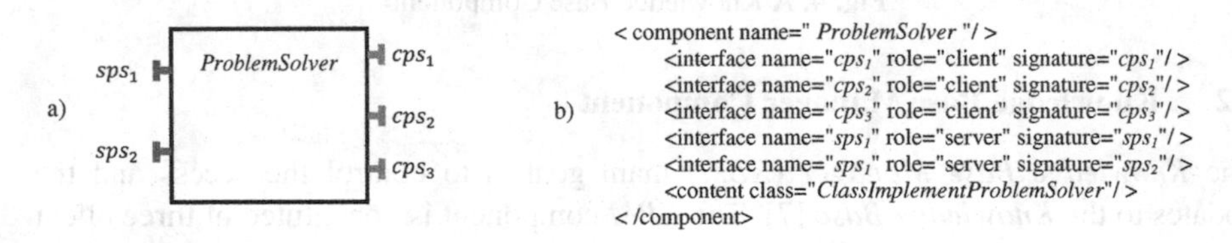

Fig. 7. A Problem Solver Component

4.5 The Behavior Engine Component

The *Behavior Engine (BE) (Expression Engine)* is responsible for analysis of the agent's current internal state and possible modifications of (parts of) that state [7]. Thus, the *BE* integrates client interfaces only that are used to collect the knowledge from the *P*, *S* and *KBM* to decide about the appropriate changes to make on the state.

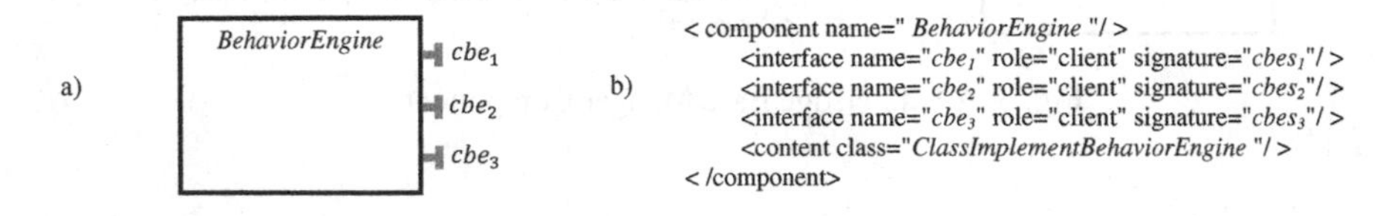

Fig. 8. A Behavior Engine Component

4.6 The State Component

State is a general abstraction for many different kinds of states a pedagogical agent can be made "aware of", as well as both volatile and persistent factors that influence the states [7]. The binding of the State component to the *Perception* and the *Action* components assures the visibility of the current pedagogical agent state.

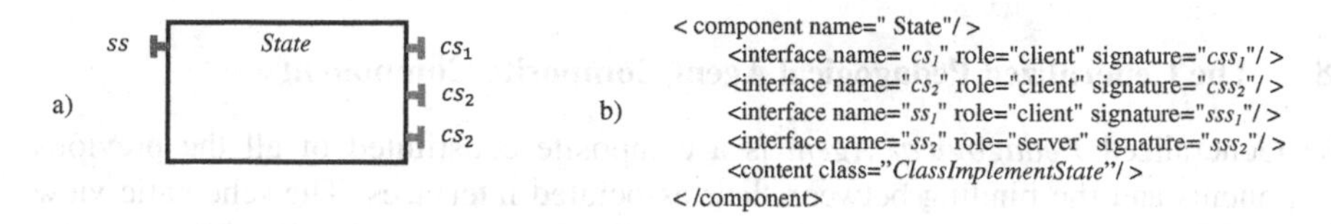

Fig. 9. A State Component

4.7 The Communication

It is responsible for perceiving the dynamic learning environment and acting upon it [7]. However, here, we subdivide the communication module to a two components as generally the case for *Agent Models*. The first is the *Action* and the second is the *Perception*. They respectively concern the sensors and the effectors of the *Pedagogical Agent*

The Perception Component
Perceptions include recognizing situations in which the pedagogical agent can intervene, such as specific student's actions, co-learner's progress, and availability of desired information [7]. It has server interfaces only to offer information about the interaction context.

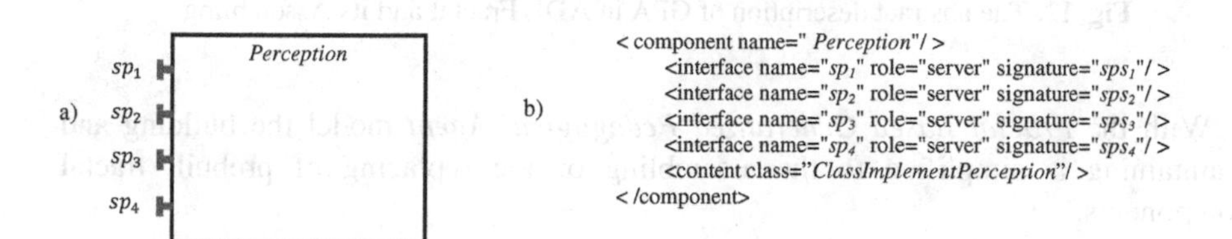

Fig. 10. A Perception Component

The Action Component
It is a primitive component that is constituted of two server interfaces. The inexistence of client interfaces is justified by its nature as entity that acts on the environment.

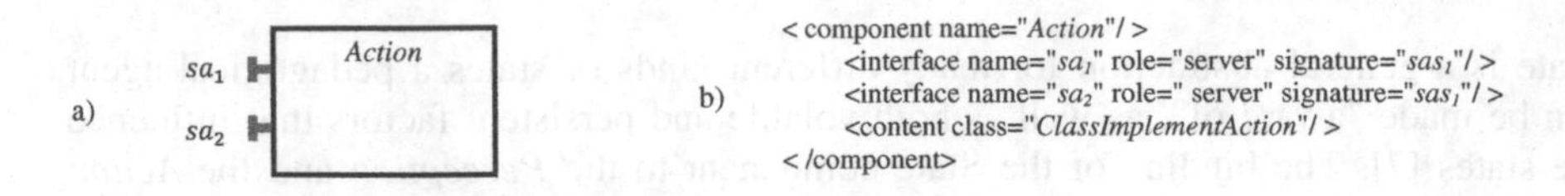

Fig. 11. An Action Component

4.8 The Generalized Pedagogical Agent Composite Component

The Generalized *Pedagogical Agent* is a composite constituted of all the previous components and the binding between their associated interfaces. The schematic view of the *ADL Fractal* abstract description of the *GPA* is represented in Fig. 12.

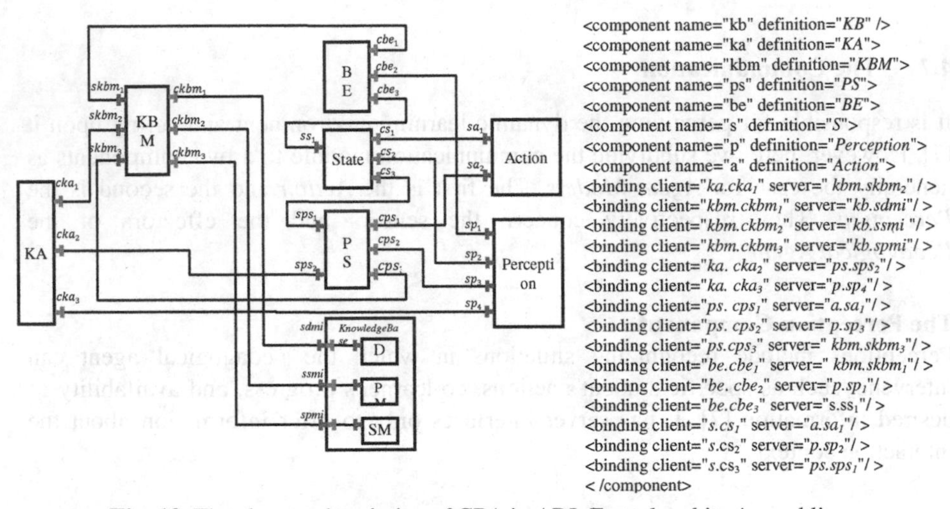

Fig. 12. The abstract description of GPA in ADL Fractal and its Assembling

With the *Fractal Based Generalized Pedagogical Agent* model the building and maintaining is simplified to the assembling or the replacing of prebuilt fractal components.

5 Discussion

The novelty of the proposed model is its use of the Fractal component to model the Generalized Pedagogical Agent. Especially, we use the *ADL Fractal* to automate the *FGPA* building and simplify its maintaining. Thus, the GPA agent is modeled as an abstract description of components and binding between them. The Implementation is achieved by the content associated to each of them. The *ADL Fractal* can be directly used by the reconfiguration primitives of the Fractal implementation, such as *Julia*, to create a concrete *FGPA*. Also, the use of the Fractal in modeling and implementing the *Generalized Pedagogical Agent* makes it suitable in cases of dynamic automatic pedagogical agent reconfiguration.

Also, because the structures of existing *Pedagogical Agents* are unknown and the implementations are not available, it is difficult to implement the whole parts and the functionalities of the *Generalized Pedagogical Agent*.

6 Conclusion

Within this paper we have presented a *Fractal Based Generalized Pedagogical Agent Model* that aims to create the Pedagogical Agent as a composite component of a set of fractal sub-components. The GPA structure, the required-provide interdependency relationships and the content of the component are described using the ADL Fractal. The assembling of the Fractal-based GPA is achieved by the reconfiguration primitives of the Fractal Model. The fundamental advantages of the proposed model are its cost-effectiveness and flexibility resulted of the clear separation between the structural parts and the separation of the abstract description in from its fulfilling. Thus, the building and the maintenance are simplified to the assembling of the pre-built approved components and the replacement of the failed components respectively. Also, the use of the component paradigm in general and the Fractal component model in particular make the porting of the Pedagogical Agent from one domain model to another highly simplified. So, the modeling of the Knowledge Domains as reusable Composite Components makes it domain independent. The same thing can be true with the other part constituting this *Pedagogical Agent*.

As a future work, we work on the integration of an adapting Engine to a framework implemented using Julia of the Pedagogical Agent. The main goal of this work is to make the *FGPA* context sensitive and auto-adaptive regarding the platforms and the environments characteristics of a ubiquitous *FGPA*.

References

1. Grondin, G., Bouraqadi, N., Vercouter, L.: MADCAR: An Abstract Model for Dynamic and Automatic (Re-)Assembling of Component-Based Applications. In: Gorton, I., Heineman, G.T., Crnković, I., Schmidt, H.W., Stafford, J.A., Ren, X.-M., Wallnau, K. (eds.) CBSE 2006. LNCS, vol. 4063, pp. 360–367. Springer, Heidelberg (2006)
2. Ketfi, N.B., Cunin, P.Y.: Adapting Applications on the Fly. In: Proceedings of the 17th IEEE International Conference on Automated Software Engineering (ASE 2002), p. 313. IEEE Computer Society, Washington, DC (2002)
3. Szyperski: Component Software - Beyond Object-Oriented Programming. Addison-Wesley/ACM Press (1998)
4. George, B., Fleurquin, R., Sadou, S., Sahraoui, H.A.: Un mécanisme de sélection de composants logiciels. L'OBJET 14(1-2), 139–163 (2008)
5. Lau, K.-K., Wang, Z.: Software component models. IEEE Trans. Software Eng. 33(10), 709–724 (2007)

6. Bruneton, T.C., Leclecq, M., Quema, V., Stefani, J.-B.: The Fractal component model and its support in Java. Software, Practice and Experience 36(11-12), 1257–1284 (2006)
7. Devedzic, V., Harrer, A.: Software patterns in its architectures. I. J. Artificial Intelligence in Education 15(2), 63–94 (2005)
8. Briot, J.-P., Meurisse, T., Peschanski, F.: Architectural design of component-based agents: A behavior-based approach. In: Bordini, R.H., Dastani, M., Dix, J., El Fallah Seghrouchni, A. (eds.) PROMAS 2006. LNCS (LNAI), vol. 4411, pp. 71–90. Springer, Heidelberg (2007)
9. Devedžić, V., Jerinić, L., Radović, D.: Communication Between the Knowledge Base and Other Modules of the Intelligent Tutoring Shell in GET-BITS Model. In: Martin-Rubio, F., Triguero-Ruiz, F. (eds.) Actas de las Segundas Jornadas de Transferencia Tecnologica de Inteligencija Artificial TTIA 1997 (Proceedings of the 2nd Conference of the Transfer Technology on Artificial Intelligence TTIA 1997), November 12-14, AEPIA Press, Malaga (1997); addenda A-1 – A-11. (R54)
10. Mullier, D., Dixon, M.: Authoring educational hypermedia using a semantic network. In: World Conference on Educational Multimedia, Hypermedia and Telecommunications, vol. 2000(1), pp. 295–300 (2000)
11. McArthur, D., Lewis, M., Bishay, M.: The Roles of Artificial Intelligence in Education: Current Progress and Future Prospects. RAND 1700 Main St. Santa Monica, CA 90407-2138
12. Lewis Johnson, W., Shaw, E.: Using Agents to Overcome Deficiencies in Web-Based Courseware. In: Proceedings of the Workshop "Intelligent Educational Systems on the World Wide Web", Kobe, Japan, August 18-22. 8th World Conference of the AIED Society (1997)
13. Roberto, V., Mea, V.D., Di, L., Conti, G.A.: MANTHA Agent based Management of Hypermedia Documents. In: Proceedings of 6th IEEE Int. Conf. on Multimedia Computing and Systems (IEEE ICMCS 1999), Firenze, vol. II, pp. 814–818. IEEE Computer Society (June 1999)
14. Virvou, M., Moundridou, M.: A Web-Based Authoring Tool for Algebra-Related Intelligent Tutoring Systems. Educational Technology & Society 3(2) (2000) ISSN 1436-4522
15. Beck, J., Stern, M., Haugsjaa, E.: Applications of AI in Education. Crossroads - Special issue on Artificial Intelligence 3(1), 11–15 (1996)
16. Frasson, C., Mengelle, T., Aïmeur, E., Gouardères, G.: An actor-based architecture for intelligent tutoring systems. In: Lesgold, A.M., Frasson, C., Gauthier, G. (eds.) ITS 1996. LNCS, vol. 1086, pp. 57–65. Springer, Heidelberg (1996)
17. Frasson, C., Mengelle, T., Aimeur, E.: Using Pedagogical Agent. In: a Multistrategic Intelligent Tutoring System.In: Workshop on Pedagogical Agents in AI-ED 1997, World Conference on Artificial Intelligence and Education, Japan, pp. 40–47 (August 1997)
18. Webber, C., Bergia, L., Pesty, S., Balacheff, N.: The BAGHERA Project: A Multi-Agent Architecture for Human Learning. In: MABLE Workshop - AIED 2001 Conference, San Antonio, TX, USA (May 19, 2001)
19. Jaques, P., Andrade, A., Jung, J., Bordini, R., Vicari, R.: Using pedagogical agents to support collaborative distance learning. In: Proceedings of the Conference on Computer Support for Collaborative Learning: Foundations for a CSCL Community, pp. 546–547. International Society of the Learning Sciences (January 2002)
20. Tecuci, G., Keeling, H.: Developing Intelligent Educational Agents with Disciple. International Journal of Artificial Intelligence in Education 10, 221–237 (1999)

21. Chou, C.-Y., Chan, T.-W., Lin, C.-J.: Redefining the learning companion: the past, present, and future of educational agents. Computers & Education 40(3), 255–269 (2003)
22. Garlan, D., Cheng, S.-W., Huang, A.-C., Schmerl, B., Steenkiste, P.: Rainbow: Architecture-Based Self-Adaptation with Reusable Infrastructure
23. Sparling, M.: "Lessons Learned through Six Years of Component Based Development" published in Castek (as published in Communications of the ACM) (April 09, 2003)
24. Virtanen, P.: Measuring and Improving Component-Based Software Development by Pentti Virtanen. University of Turku, Department of Computer Science, FIN-20014 Turku Finland (2003)
25. Johnson, W.L., Rickel, J.W., Lester, J.C.: Animated pedagogical agents: Face-to-face interaction in interactive learning environments. International Journal of Artificial Intelligence in Education 11(1), 47–78 (2000)
26. Bruneton, E., Coupaye, T., Leclercq, M., Quema, V., Stefani, J.-B.: An open component model and its support in java. In: Crnković, I., Stafford, J.A., Schmidt, H.W., Wallnau, K. (eds.) CBSE 2004. LNCS, vol. 3054, pp. 7–22. Springer, Heidelberg (2004)
27. Lau, K.K., Wang, Z.: A survey of software component models. In: 31st EUROMICRO Conference on Software Engineering and Advanced Applications. IEEE Computer Society (2005)
28. Bruneton, E., Coupaye, T., Stefani, J.B.: Recursive and dynamic software composition with sharing. In: Proceedings of the 7th ECOOP International Workshop on Component-Oriented Programming (WCOP 2002), pp. 133–147 (June 2002)
29. Emmerich, W., Kaveh, N.: Component technologies: Java beans, COM, CORBA, RMI, EJB and the CORBA component model. In: Proceedings of the 24th International Conference on Software Engineering, ICSE 2002, pp. 691–692. IEEE (May 2002)
30. Clements, P.C.: A survey of architecture description languages. In: Proceedings of the 8th International Workshop on Software Specification and Design. IEEE Computer Society (March 1996)
31. Medvidovic, N., Taylor, R.N.: A classification and comparison framework for software architecture description languages. IEEE Transactions on Software Engineering 26(1), 70–93 (2000)

Partially Centralized Hierarchical Plans Merging

Said Brahimi[1,2], Ramdane Maamri[2], and Zaidi Sahnoun[2]

[1] University of Guelma, BP 401, Guelma 24000, Algeria
brahimi.said@yahoo.fr
[2] LIRE Laboratory, University of Constantine 2, BP 325 Constantine 25017, Algeria
{rmaamri,sahnounz}@yahoo.fr

Abstract. In multi-agent systems evolving in complex and dynamic environments, it is difficult to ensure an efficient coordination of agents plans. To deal with this issue searchers have exploited the concept of the abstraction for defining adequate plan representations, and also for providing algorithms serving to planning, coordination, and execution. By ignoring some less important details, a solution can be found more simply. In this paper we provide an adequate plan representation model, by taking advantage of hierarchical plans and the ability to reason on abstract tasks. We use the obtained model to provide some principles for implementing Partially Centralized scheme for Hierarchical Plans merging. Based on summary information and the ability to localize the interference between tasks supported by our model, the global plan is dynamically decomposed to a set of partial-global-plan that may be analysed in decentralized manner. This hybridization, between centralized and distributed coordination, can favour the interleaving of planning and execution, some part of global plan can be repaired when others are in execution.

Keywords: Hierarchical plan merging, Multi-agent Tasks organization, Partially Centralized coordination, Planning-Execution interleaving.

1 Introduction

In multi-agent systems evolving in complex and dynamic environments, it is difficult to ensure an efficient coordination of agents plans. To deal with this issue searchers have exploited the concept of the abstraction for defining adequate plan representations, and also for providing algorithms serving to planning, coordination, and execution. By ignoring some less important details, a solution can be found more easily [11][4].

Many works [7][8][5][6][9] have adopted a plans models having hierarchical structure, each level in the hierarchy represents an abstraction level. The highest level includes compound (or abstract) tasks and the lower level includes primitive (or elementary) tasks. The intermediate levels include less abstract tasks. Links between different levels represent abstract tasks decomposition (or refinement). Other approaches such as [3][4][12][10] have proposed an extended model of hierarchical plans by introducing so called summary information. Annotating the

A. Bădică et al. (eds.), *Recent Developments in Computational Collective Intelligence*,
Studies in Computational Intelligence 513,
DOI: 10.1007/978-3-319-01787-7_6,

abstract tasks, this information summarizes pre, post conditions, and resources needed of tasks in lower levels. The use of these extended models allows to reason on abstract tasks in order to deal with interdependency of concurrent tasks. This ability can aide to reduce the complexity of solution finding and to make possible the interleaving of local planning and execution.

Despite positive aspects that characterize the used model, these approaches remain not suitable to address the dynamic aspect of tasks and plans, especially where coordination must be interleaved with the execution. To deal with this insufficiency we defined in previous work [2] a model of hierarchical plans, called HPlNet, that takes into account the representation of flexible plans and offers the necessary features allowing to monitor agents' plans in all its states. However, this model is not suitable to deal with synchronisation between tasks.

In this paper we provide an adequate plan representation model, by taking advantage of hierarchical plans and the ability to reason on abstract tasks. We use the obtained model to provide some principles for implementing Partially Centralized scheme for Hierarchical Plans merging. Based on summary information and the ability to localize the interference between tasks supported by our model, the global plan is dynamically decomposed to set of partial-global-plan that may be analysed in decentralized manner. This hybridization, between centralized and distributed coordination, can favour the interleaving of planning and execution, some part of global plan can be repaired when others are in execution. Based on this idea, we are expected to generalise our previous [1] by using the proposed model.

There are another works [9][6] similar to ours work. They are based on the idea of delegating a part of planning problem of parent agent to its child agents. The multi-agent planning scheme is relatively similar to the one in our work. The main difference is situated in the splitting method. The work was not taken in consideration the independent splitting. This point is important especially in the case of dynamic planning.

The remaining of this paper is organized as follows. In the second section we outline the formalism used to represent agents' plans. In the section 3 we provide the principle ideas of the centralized scheme permitting to merge agents' plans. In the section 4 we explain the key idea related to the identification of independent Sub-Problems and the way that can be exploited to obtain coordination mechanism that is partially centralized. Finally, we conclude the paper by indicating the positive aspects and some perspectives related to future works.

2 Plans Representation Concept and Analysis

In this section we present the formalism model used to represent agents' plans which are assumed represented by a set of tasks hierarchically organized. Firstly, we begin by presenting our previous model HPlNet (Hierarchical Plan Net), proposed in [2]. Secondly we explained how this model is extended in order to handle plans synchronization.

2.1 Hierarchical-Plan-Net (HPlNet)

Definition 1. *A Hierarchical-Plan-Net is defined by the tuple:* (P, T, F, s, e, Res, R) *where:* P *is a finite set of places;* $T = T_{elem} \cup T_{abs}$ *is a finite set of transitions, disjoint union of elementary,* T_{elem}*, and abstract,* T_{abs} *transitions,* T_{abs} *may be empty;* $F \subseteq P \times T \cup T \times P$*;* s *is a particular place representing the source place* ($\bullet s = \emptyset$)*;* e *is a particular transition representing thc cnd transition* ($e\bullet = \emptyset$)*;* $Res : T \rightarrow \Re q \times \Re q$ *is a function defining the sets of resources to consume and to produce of each task (transition),* $\Re q = \{(rs_1, q_1), ..., (rs_n, q_n)\}$ *where* rs_i *is resource name and* $q_i = (x, y)$ *represents the lower (*x *or* q_i^-*) and upper (*y *or* q_i^+*) quantity (or number) of* rs_i*; The set of all resources is denoted by* $\Re$*;* $R \subseteq T_{abs} \times HPlNet$*,* $R = \cup_{t \in T_{abs}} R_t$*, is a finite set of refinement rules for all abstract transitions(must be empty if* $T_{abs} = \emptyset$*), each rules* $r \in R$ *associates to a transition* $t \in T_{abs}$ *a refinement* $HPlNet$*.* R_t *denotes the set of all rules used to refine the task* t*;* (P, T, F) *is a petri net to have either of the following two structures:*

- *All transitions must be executed in single sequence. i.e.* $P = \{s\} \cup \cup_{i=1}^{m} \{p_i\}$*,* $T = \cup_{i=1}^{m} \{t_i\} \cup \{e\}$*, and* $F = \{(s, t_1)\} \cup \cup_{i=1}^{m} \{(t_i, p_i)\} \cup \cup_{i=1}^{m-1} \{(p_i, t_{i+1})\} \cup \{(p_m, e)\}$*. in this case,* (P, T, F, s, e) *is called Sequential-TaskPN node*
- *All transitions (except fork and end-transitions) can be concurrently executed, i.e.* $P = \cup_{i=1}^{m} \{p_i, p_i'\} \cup \{s\}$*,* $T = \cup_{i=1}^{m} \{t_i\} \cup \{f, e\}$*,* $F = \{(s, f)\} \cup \cup_{i=1}^{m} \{(f, p_i)\} \cup \cup_{i=1}^{m} \{(p_i, t_i)\} \cup \cup_{i=1}^{m} \{(t_i, p_i')\} \cup \cup_{i=1}^{m} \{(p_i', e)\}$*, the tasks are connected to source place by a fork transition* f*. in this case,* (P, T, F, s, e) *is called Parallel-TaskPN node*

A HPlNet $\wp = (P, T, F, s, e, Res, R)$ may be considered as a tree of nodes. The root of this tree is (P, T, F, s, e) where $T_{abs} = \{t_0\}$ and $T = e, t_0$; t_0 is highest level task of $\wp$. The leaves of the tree are nodes where $R = \emptyset$ ($T = T_{elem}$). The intermediate nodes are characterized by $R \neq \emptyset$ ($T_{abs} \neq \emptyset$). Abstract transitions model abstract (or compound) tasks and elementary transitions model atomic (elementary) tasks.

The execution of plan is modelled by state evolution (dynamic aspect) of HPlNet. Like the case of petri net, dynamic aspect of HPlNet is represented by the evolution of (extended) marking states.

Definition 2. *An extended marking of HPlNet* $\wp$ *is defined by* (Tr, Cxt)*, where (i)* $Tr = (N, n_0)$ *is a tree such that* N *is the set of node; each node* n *is a tuple* (Pl, Mp, Mt) *such that* Pl *is a node in* $\wp$*;* $Mp : P(Pl) \rightarrow \{0, 1\}$*, and* $Mt : T_{abs}(Pl) \rightarrow N \cup \varepsilon$ *(*ε *denotes the absence of token) is a marking function of abstract transitions;* $n_0 \in N$ *is the root of tree; a node* n' *is the child of* n *in* Tr *iff* $\exists t \in T_{abs}(Pl(n))$ *such that* $(t, Pl(n')) \in R(Pl(n))$*, and* $Mt(n)(t) = n'$ *(ii)* $Cxt = \{(rs_1, qt_1), ..., (rs_n, qt_n)\}$ *is a finite set of available resources, where* qt_i *is the quantity of resource* rs_i*. We write* $qt_{Cxt}(rs)$ *to denote the quantity of* rs *in* Cxt*.*

The initial extended marking is (Tr_0, Cxt_0) such that $Tr_0 = (\{n_0\}, n_0)$ where $n_0 = (Pl, Mp_0, Mt_0)$, $Mp_0 = s(Pl(n_0))$, Mt_0 is the initial marking where

$Mt(t_0) = \varepsilon$; the final marking (Tr_f, Cxt_f) such that Tr_f is an empty tree (that has no node). A step sp between two marking states (Tr, Cxt) and (Tr', Cxt'), denoted by $(Tr, Cxt) \xrightarrow{sp} (Tr', Cxt')$, concern the firing of elementary transition, end-transition, or (selected) refinement-rule.The firing condition and rules are explained in detail in [2].

The advantage of HPlNet is to allow the reasoning in the abstraction level based on the summary information about the resources. We can decide with certainty that the safe execution of the plan is *sure* or *impossible*[2]. However, the model does not take into account the synchronization between plans. In the next subsection we provide an extension of this model to overcome this limitation.

2.2 HPlNet with Synchronization: SHPlNet

Definition 3. *A Hierarchical-Plan-Net with Synchronization (SHPlNet) is defined by the tuple:* $(P, T, F, s, e, Res, R, S)$ *where:* (P, T, F, s, e, Res, R) *is HPlNet;* $S = (\cup_i P_i^s, \cup_i T_i^s, \cup_i W_i^s, \cup_i F_i^s)$ *is a Petri net such that* $(P_i^s, T_i^s, W_i^s, F_i^s)$ *can have either of the following structure:* $(\{p, rs\}, \{t\}, \{(t, p), (t, rs)\}, \{((t, p), 1), ((t, rs), x)\})$ *to represent a production of quantity* x *of the resource* $rs \in \Re$, $(\{p, rs\}, \{t\}, \{(p, t), (rs, t)\}, \{((p, t), 1), ((rs, t), x)\})$ *to represent a consumption of quantity* x *of the resource* $rs \in \Re$, $(\{p\}, \{t_1, t_2\}, \{(t_1, p), (p, t_2)\}, \{((t_1, p), 1), ((p, t_2), 1)\})$ *to represent an ordering between tasks,* $(\{p\}, \{r\}, \{(p, r)\}, \{((p, r), 1)\})$ *to represent an inhibition of the rule* r*. Each transition in* T^s *is a transition or a rule defined in HPlNet.*

The Petri net S defined in the definition 3 constitutes a coordination module including synchronization constraints that enforce an execution order and make a resource exchange between concurrent tasks. It includes also constraints that enforce the selection of one refinement rule.

The state of plan p represented by SHPlNet is modeled by the extended marking of SHPlNet (definition 4) that takes into account marking state of synchronisation net.

Definition 4. *An extended marking of SHPlNet* $\wp = (P, T, F, s, e, Res, R, S)$ *is defined by the tuple* (Tr, Cxt, M_s) *such that* (Tr, Cxt) *is the extended marking of HPlNet* (P, T, F, s, e, Res, R)*; and* M_s *is a marking of* S*, the initial and final marking of* S *is* $M_{s_0} = 0$.

The steps of SHPlNet are either a firing of elementary-transition, end-transition, or refinement-rule. The definition 5 and 6 below state respectively the firing condition and firing rules.

Definition 5. *Given an extended marking* (Tr, Cxt, M_s)*, a node* n *in* Tr*, a step* sp *is enabled in* (Tr, Cxt, M_s)*, denoted by* $(Tr, Cxt, M_s) \xrightarrow{sp}$*, iff: (i)* $(Tr, Cxt) \xrightarrow{sp}$*; (ii)* $sp \in R_t(n) \wedge t \in T_s(M_s) \Rightarrow M_s \xrightarrow{t}$*; (iii)* $sp \in T_s \Rightarrow M_s \xrightarrow{sp}$

In the definition 5, the condition (ii) states that all step appearing in S cannot be enabled if it is not enabled in M_s. The condition (iii) states that an abstract transition appearing in S cannot be refined if it is not enabled in M_s.

Definition 6. *Let (Tr, Cxt, M_s) be an extended marking, n be a node in Tr, the firing of a step sp leads to the extended marking (Tr', Cxt', M_s'), denoted by $(Tr, Cxt, M_s) \xrightarrow{sp} (Tr', Cxt', M_s')$, if and only iff:*

case 1: $sp \in T_{elem}(Pl(n))$: $(Tr, Cxt) \xrightarrow{sp} (Tr', Cxt')$; $sp \in T_s \Rightarrow M_s \xrightarrow{sp} M_s'$; *and* $sp \notin T_s \Rightarrow M_s' = M_s$

case 2: $sp \in R_t(Pl(n))$: $(Tr, Cxt) \xrightarrow{sp} (Tr', Cxt')$; $t \in T_s \Rightarrow M_s' = M_s - W_s(.,t)$; *and* $t \notin T_s \Rightarrow M_s' = M_s$

case 3: $sp = e(Pl(n))$ *and* n *is child of* n' *by* $r = (t', Pl(n))$: $(Tr, Cxt) \xrightarrow{sp} (Tr', Cxt')$; $t' \in T_s \Rightarrow M_s' = M_s + W_s(t', .)$; *and* $t' \notin T_s \Rightarrow M_s' = M_s$

in the definition 6 the case 1 states that the firing on an elementary transition appearing in S leads to its firing in S. The case 2 states that the firing of refinement rule of a transition that appear in S leads to the consumption of tokens. The production of the tokens will be after the firing of end transition of the node refining the abstract transition (case 3).

In the same way as a plan represented by HPlNet, a plan represented by SHPlNet can be analysed from abstract level. Based on the summary information about the resources associated to transitions, we can distinguish certainty a cases in which the safe execution of the plan in some situation is sure or impossible.

Given a SHPlNet $\wp = (P, T, F, s, e, Res, R, S)$ the safe execution of $\wp$ in the state (Tr, Cxt, M_s) is

- *Sure*: if the safe execution of (P, T, F, s, e, Res, R) is sure in the state (Tr, Cxt) and there is no cycle in S.
- *Impossible*: if the safe execution of (P, T, F, s, e, Res, R) is impossible in the state (Tr, Cxt) or there is a cycle in S

Cycles in synchronisation net S prevents the execution of any task of $\wp$ appearing in S.

A plan whose execution is impossible is a plan that has no way to ensure the success of plan execution. A plan whose safe execution is sure is a flexible plan. It can be executed correctly regardless of the choice to be taken to refine abstract transitions and the execution order of concurrent tasks. Between these two cases there is an uncertainty case in which the success of execution may be possible. To address this incertitude, the plan must be reorganised by updating the synchronisation net (block some refinement rules and/or add some constraints on the execution of tasks).

In the next section we provide a centralized technic for merge a set of agents' plans in safe multi-agent plan.

3 Centralized Merge of Agents Plans

Safe merging of set of plans $\{\wp_i\}$ of agents $\{A_i\}$ consist of integrating this plans in a multi-agent plan, $G\wp$, whose safe execution is sure. The first step of plans merge is to combine the tasks performed by different agents in single SHPlNet. Formally, the link between multi-agent plan and agent plan is as follow:

Let $\wp_i = (\{s_i, p_i\}, \{t_i, e_i\}, \{(s_i, t_i), (t_i, p_i), (p_i, e_i)\}, s_i, e_i, Res_i, R_i, S_i)$ be a SHPlNet representing the plan of agent A_i such that $i \in \{1..n\}$, the multi-agent plan is represented also by a SHPlNet $G\wp = (\{s, p\}, \{t, e\}, \{(s, t), (t, p), (p, e)\}, s, e, Res, \{(t, pl)\}, S)$ such that: $pl = (P, T, F, s', e', Res', R')$, $P = \{s'\} \cup \cup_{i=1}^{n} \{s_i, p_i\}$, $T = \cup_{i=1}^{n}\{t_i\} \cup \{e', f'\}$, $F = \{(s', f')\} \cup \cup_{i=1}^{n}\{(f', s_i)\} \cup \cup_{i=1}^{n}\{(s_i, t_i), (t_i, p_i)\} \cup \cup_{i=1}^{n}\{(p_i, e')\}$, $S = (\cup_{i=1}^{n} P_s(S_i), \cup_{i=1}^{n} T_s(S_i), \cup_{i=1}^{n} W_s(S_i), \cup_{i=1}^{n} F_s(S_i))$, $R' = \cup_{i=1}^{n} R_i$. The Petri net S represents intra-agent synchronization constraints (if it is established between the tasks of same agent) and inter-agent (if it is established between the tasks of different agents). Inter-agent synchronization constraints represent a coordination structure between agents.

The plans' merge consists of analysing the SHPlNet representing $G\wp$ from top level to ground level. At each level of abstraction synchronization constraints between the tasks of agents' plans may be imposed, and some alternatives may be favored in order to minimize the consumption of resources required by $G\wp$.

To reduce the computational complexity, we'll only focus on the required resources whose quantity is insufficient for carrying out the task t_0 of SHPlNet representing $G\wp$, noted Critical-Resource (CR), and the tasks related to these resources, noted Critical-Task (CT). Formally a resource rs is critical for the execution of $G\wp$ in the context Cxt iff $qt_{Cxt}(rs) < q^+$ such that $(rs, q) \in Res(t_0).cons$. In this case, a safe multi-agent plan $G\wp$ can be redefined as one whose t_0 is not Critical-Task (i.e. requires no critical resource). We define also so called Analysis Tree in which the critical-tasks and critical-resources are distinguished. Formally,it is defined by $AnTr = (NN, nn_0, child)$, where NN is a finite set of nodes and $nn_0 \in NN$ is the root of tree. Each node nn is a tuple $(n, Ctsk, Crs)$ such that n is a node in $G\wp$, $Ctsk \neq \emptyset$, and $Crs \neq \emptyset$ are respectively finite set of the critical tasks and critical resources; $child$ is a set of (nn, r, nn') where $nn = (n, Ctsk, Crs)$, $nn' = (n', Ctsk', Crs')$, $r = (t, pl)$ where $t \in T(Pl(n))$ and $pl = Pl(n')$.

The merge procedure search in the space of global plan according to the strategy of consumption minimization and production maximization the of all critical resources in order to obtain a global plan that has no critical task. In the two case of this strategy, the following decisions can be taken:

- Refinement rule selection: Choose one refinement rule (The other must be blocked) from those can be used to refine an abstract task (transition). Let $t \in T_{abs}(n)$ be a transition in node n in $G\wp$, the selection of the rule $r = (t, pl) \in R_t(n)$ implies updating $S = (P^s, T^s, F^s, W^s)$ as follow: $P^s \leftarrow P^s \cup \{p/p \text{ is new place}\}$, $T^s \leftarrow T^s \cup \{r\}$, $F^s \leftarrow F^s \cup \{(p, t)\}$, $W^s \leftarrow \{((p, t), 1)\}$
- Tasks ordering: Impose a sequential execution between concurrent tasks (transitions). It is generally used between tasks belonging to the same node. Let $t_1, t_2 \in T(n)$ be two concurrent transitions in node n in $G\wp$, the ordering of t_1 and t_2 implies updating $S = (P^s, T^s, F^s, W^s)$ as follow: $P^s \leftarrow$

$P^s \cup \{p/p \text{ is new place}\}$, $T^s \leftarrow T^s \cup \{t_1, t_2\}$, $F^s \leftarrow F^s \cup \{(t_1,p),(p,t_2)\}$, $W^s \leftarrow W^s \cup \{((t_1,p),1),((p,t_2),1)\}$

- Addition of producer/consumer link between two tasks (who produces or releases a resources and who uses or consumes these resources). Let $t_1, t_2 \in T(n)$ be two concurrent transitions belonging respectively to nodes n_1, n_2 in $G_\wp$, the addition of producer/consumer link between t_1 and t_2, related to a quantity qt of resource rs, implies updating the synchronization $S = (P^s, T^s, F^s, W^s)$ as follow: $P^s \leftarrow P^s \cup \{p, rs/p \text{ is new place}\}$, $T^s \leftarrow T^s \cup \{t_1, t_2\}$, $F^s \leftarrow F^s \cup \{(t_1,p),(t_1,rs),(p,t_2),(rs,t_2)\}$, $W^s \leftarrow W^s \cup \{((t_1,p),1), ((t_2,rs),qt),((p,t_2),1),((rs,t_2),qt)\}$

The generic procedure $Merge$ trys to refine the initial global plan where the set of agents' plans are arrenged without synchronisation. The input of the procedure is hence the initial analysis tree, $AnTr$ and the output is a set of analysis tree representing the possible solutions, *solutions*. The global plan refinement is made by applying one of the three operators discussed above, Refinement rule selection, Tasks ordering, and Addition of producer/consumer link between tasks. Note that the selection of one operator and the strategy of its applying are implemented in the functions *choose* and $decisionMaking$.

```
1: procedure MERGE(AnTr)
2:     cand ← {AnTr};
3:     solutions ← ∅;
4:     while true do
5:         if cand = ∅ then
6:             return solutions ;
7:         end if
8:         AnTr' ← pop(cand);
9:         if Crs(nno(AnTr')) = ∅ then
10:            solutions ← solutions ∪ {AnTr'};
11:        end if
12:        if safeEexecution(AnTr') = possible then
13:            decision ← choose({ruleSelection, orderinAdding, prodConsAdding});
14:            cand ← cand ∪ decisionMaking(decision, AnTr');
15:        end if
16:    end while
17:    return solutions
18: end procedure
```

This algorithm may be simply implemented in a single coordination agent that collects all plans agents in order to maintain a global consistency between them. Tasks of agents may be related to disjoint subset of critical-resources. In this case it will be beneficial to treat these tasks independently and in decentralized way. In the next section we deal with this idea. We will explain how decomposing de merge problem and how monitoring partially centralized coordination.

4 Partially Centralized Plans Merging

Reasoning on abstracts plans (using summary information about resources) allows to characterize the central aspect of how to identify and to handle the threat relationships between tasks in abstract level of plans. We exploit this principle to provide a technique that help to identify the fact that some subset of abstracts tasks can (or cannot) be independently analyzed. Hence, analysis of a global-plan can lead to independently analyze several partial-global-plans. The simple example concerns the case, where the set of tasks in global plan is split to disjoint sub set of tasks which are related to disjoint sub set of resources. Identification of the subset of interfered tasks (those concerned by conflict removing) can aid to divide merge problem into set of sub problems which can be solved by several coordination-agents in parallel manner.

As it is described in our previous work [1], the identification of independents-sub-problems can be embodied in the centralized coordination process as intermediate step. In each (or in some) analysis iteration, the (partially) global-plan is analysed and examined to be divided. The independent sub-problems may be assigned to several coordination agents.

The problem dividing is based on groping the tasks related to shared resources in one set. In this fashion the treatment of interfered tasks has to dual only with concerned tasks (and therefore between concerned agents). To facilitate the identification of such sub-problem, the set of elementary tasks or those that have not a refinement in the analysis tree $AnTr$ are denoted by the $LTsk(AnTr)$.

Formally independent sub-problem of $AnTr = (NN, nn_0, child)$ is defined by $SubInd(AnTr) = \{AnTr_i / i \in \{1, k\}\}$ Such that:

a) $LTsk(AnTr_i) \subset LTsk(AnTr)$, $\cup_{i=1}^{k} LTsk(AnTr_i) = LTsk(AnTr)$, and $\cap_{i=1}^{k} LTsk(AnTr_i) = \emptyset$
b) $\forall i, j \in \{1, k\}, i \neq j, \forall t \in LTsk(AnTr_i), \forall t' \in LTsk(AnTr_j)$, t and t' are not related to same critical resources
c) $\forall i \in \{1, k\}, \forall t, t' \in LTsk(AnTr_i)$, t and t' are related (directly or by transitivity) to same critical resources

The condition a) characterizes the disjoint union of independent sub-problem. The condition b) characterizes *Inter-Independency* of tasks belonging to different sub problems. The condition c) characterizes *Intra-dependency* of tasks in one sub-problem.

The independent sub-problems refinement may be assigned to different coordination-agents and so on. The result is partially centralized scheme of coordination. To implement this idea our previous work [1] can provide more explication.

5 Conclusion

In this paper we firstly provided formalism called SHPlNet (Hierarchical Plan Net with synchronisation) by extending our previous model HPlNet proposed in

[2]. We are focused on the distinction between hierarchical tasks representation and synchronization ways. We made this extension to be able taking into account the representation of concurrent tasks synchronization and interaction characterizing multi-agent plans. Starting from this ability and the ability to reason on the abstract tasks (or plans), we proposed a partially centralized scheme that allows to coordinate a set of hierarchical plans in multi-agent context.

The main idea consists of incorporation of problem splitting technique into centralized coordination. Based on the summary information associated to abstract tasks, the agents' plans can be analyzed and the interference between tasks can be localized in independent sub set. Therefore the global plan can dynamically be decomposed to set of partial-global-plan that may be analyzed in decentralized manner. This hybridization, between centralized and distributed coordination, is appropriate especially in complex and dynamic environments. This scheme of coordination can favor the interleaving of planning and execution, some part of global plan can be repaired when others are in execution.

In The future work will be focused on the expansion of coordination mechanism. We will deal especial with the monitoring the hierarchical coordination process in the case of dynamic planning (where the planning and execution process are interleaved).

References

1. Brahimi, S., Maamri, R., Sahnoun, Z.: Hierarchical coordination - towards scheme based on problem splitting. In: Filipe, J., Fred, A.L.N., Sharp, B. (eds.) ICAART 2010 - Proceedings of the International Conference on Agents and Artificial Intelligence, vol. 2, pp. 327–331. INSTICC Press (2010)
2. Brahimi, S., Maamri, R., Sahnoun, Z.: Model of petri net for representing an agent plan in dynamic environment. In: O'Shea, J., Nguyen, N.T., Crockett, K., Howlett, R.J., Jain, L.C. (eds.) KES-AMSTA 2011. LNCS, vol. 6682, pp. 113–122. Springer, Heidelberg (2011)
3. Clement, B.J., Durfee, E.H.: Theory for coordinating concurrent hierarchical planning agents using summary information. In: Proceedings of the Sixteenth National Conference on Artificial Intelligence and the Eleventh Innovative Applications of Artificial Intelligence, AAAI 1999, pp. 495–502. American Association for Artificial Intelligence, Menlo Park (1999)
4. Clement, B.J., Durfee, E.H., Barrett, A.C.: Abstract reasoning for planning and coordination. Journal of Ai Research 28, 453–515 (2007)
5. Corkill, D.D.: Hierarchical planning in a distributed environment. In: Proceedings of the 6th International Joint Conference on Artificial Intelligence, IJCAI 1979, vol. 1, pp. 168–175. Morgan Kaufmann Publishers Inc., San Francisco (1979)
6. desJardins, M., Wolverton, M.: Coordinating planning activity and information flow in a distributed planning system. AI Magazine 20(4) (1999) (Winter)
7. Erol, K., Hendler, J., Nau, D.: Semantics for hierarchical task network planning. Tech. rep. CS-TR-3239, University of Maryland (1994)
8. Erol, K., Hendler, J.A., Nau, D.S.: Umcp: A sound and complete procedure for hierarchical task-network planning. In: Hammond, K.J. (ed.) AIPS, pp. 249–254. AAAI (1994)

9. Hayashi, H.: Stratified multi-agent HTN planning in dynamic environments. In: Nguyen, N.T., Grzech, A., Howlett, R.J., Jain, L.C. (eds.) KES-AMSTA 2007. LNCS (LNAI), vol. 4496, pp. 189–198. Springer, Heidelberg (2007)
10. Lotem, A., Nau, D.S.: New advances in graphhtn: Identifying independent subproblems in large htn domains. In: Chien, S., Kambhampati, S., Knoblock, C.A. (eds.) AIPS, pp. 206–215. AAAI (2000)
11. Sacerdott, E.D.: Planning in a hierarchy of abstraction spaces. In: Proceedings of the 3rd International Joint Conference on Artificial Intelligence, IJCAI 1973, pp. 412–422. Morgan Kaufmann Publishers Inc., San Francisco (1973)
12. Thangarajah, J., Padgham, L., Winikoff, M.: Detecting and avoiding interference between goals in intelligent agents. In: Proceedings of the International Joint Conference on Artificial Intelligence (IJCAI), pp. 721–726. Academic Press (2003)

An Agent-Based Simulation of Christakis-Fowler Social Model

Antonio Gonzalez-Pardo, Raul Cajias, and David Camacho

Computer Science Department,
Escuela Politécnica Superior,
Universidad Autónoma de Madrid
{antonio.gonzalez,david.camacho}@uam.es, raul.cajias@gmail.com

Abstract. We propose an agent-based simulation system that allows to design and test social models and also to persist the state of the social network during the execution. Using graph-based community identification and tracking algorithms, the network evolution can be analysed to characterize and compare model implementations. This system has been tested using the Christakis-Fowler model. A description of the system is given, as well as experimental results obtained from this model.

Keywords: Agent-Based Social Simulations, Complex Network Metrics, Social Model Analysis, Community-Finding Algorithms.

1 Introduction

The wide acceptance of social on-line technologies has generated a wealth of data documenting interactions between real people in real settings. Those familiar with data mining methodologies can easily capture and study the dynamics of social networks by exploiting one of the growing numbers of on-line social data sources. Currently, available data sources span almost every aspect of social life, from personal relationships [1], professional life [2], opinion and information networks, location-based relationships [3] and group-work/trust networks [4], making the field of social research ripe for a new wave of research based on real-world data.

When studying on-line social networks (OSNs), correct sampling methodologies and characteristics of the sample data obtained are often the main focus of published works. Beyond these valuable contributions however, a model that may yield the network characteristics observed is often not included in the studies – citing instead *small-world* and *free-scale* properties [5] were detected – as a possible clue of the link-generation process that may be at play in the network.

This paper present an agent based social simulator that tries to determine whether the agents show a social behaviour or not. In order to do that, a well known social model (called *Christakis-Fowler*) has been implemented and the

A. Bădică et al. (eds.), *Recent Developments in Computational Collective Intelligence*,
Studies in Computational Intelligence 513,
DOI: 10.1007/978-3-319-01787-7_7, © Springer International Publishing Switzerland 2014

result obtained from the simulator has been compared with the ones observed in real models in [6].

The following sections explains the three fundamental pillars of this work which are 'Agent-based social simulation', 'Complex Networks' and 'Community identification'. Then we describe the social model used in this work and the experimental results.

2 Agent-Based Social Simulation

Agent-based social simulations (ABSS) represent a major modelling approach for social simulations. The first agent based social model developed represented people as agents, and their interactions as social processes [7]. In later years, technological advancements have made large-scale simulations of models for entire societies [8]. Agents and agent-based computing have been widely cited in the literature as a scalable and straight-forward methodology for dealing with complex systems [9].

In a social system, the network of social interactions shifts and evolves over time as a result of the aggregation of the many micro-decisions taken by social agents that adapt and respond to the behaviour of their peers. Statistical mathematical methods that measure the network as a whole can be very useful in characterizing it, but at the same time fail to arrive at a model for how agent interaction aggregates towards emergence. Defining the system as a function of agent behaviour can help tackle this complexity [10].

The field of ABSS borrows from the domains of *agent-based computing*, computer simulations and the social sciences to provide a framework of social simulation based on agent modelling and artificial intelligence [11]. The result is a methodological approach to the testing, refinement and extension of theories, yielding a deeper understanding of fundamental causal mechanisms in multi-agent systems whose study is currently separated by artificial disciplinary boundaries [10].

Agent-based models are characterized by their flexibility: they can incorporate a wide variety of actor-driven micro-mechanisms influencing tie-formation. They also allow the parametrization of social influences to study their effect on the network as a whole [12]. While the social sciences have a long history of employing computer simulations, agent-based simulations have only been partially developed [11]. To that effect, this paper presents a system to explore the current state of the art of agent-based social models using new metrics and to provide an environment in which to develop new models.

3 Complex Networks

Complex network theory aims to provide a formal mathematical framework to work with systems of multiple and interdependent components. To do so, much attention is given to the topological structure of the component dependency network. As a result, a number of important graph-based metrics have been proposed and are currently used to characterize real-world networks.

Costa [13] provides a comprehensive review of commonly used complex system metrics, which include:

- **Node** and **edge degree**. In network theory nodes are the elements that conform the networks, while edges are the connections between nodes. The simplest metrics consist on analysing the mean number of edges per nodes.
- **Distance metrics**, measuring the distance between two given nodes. Some examples include the *characteristic path length*, the *shortest path* and *geodesic path*.
- **Cluster** and **cycle** indexes, which measure the number of three-way connections and number of cyclic paths in a graph, respectively.
- Node and edge **centrality metrics** which quantify how well connected a node is, or how well an edge connects members nodes in a graph, respectively.
- **Community identification** algorithms, which spot nodes that share more edges amongst themselves than to other nodes in a graph.

Topological characteristics of a social graph have been used before to generate dynamic models. In [5], Barabasi used the *cluster percolation method* (CPM) proposed by Palla et. al. [14] to develop a social evolution model, based on the behaviour of communities in longitudinal social datasets like phone-call records and academic citations networks. The assortativity[1] of nodes in a graph has also been explored in developing models like scale free network [15,16] as well as a number of other important models, suggesting the great contribution complex network metrics have been to the study of social network dynamics.

4 Community Identification

Algorithms that find communities of nodes and track them over consecutive snapshots of a dynamic graph are of special importance to the development of the simulation system. These methods identify clusters of nodes that share higher density of links with each other (inter-community) than to other nodes in the network [17] and may serve to characterize the network structure at a higher level than node/edge degree distribution. Indeed, the work by Yin et. al. [6] suggests that nodes/edge degree metric – such as the ones used in [6] and [18] to propose *small-world* and *free-scale* properties – may miss out on high-order structures found in real-world social networks. In [17], Fortunato presents a quite thorough review of the state of art on the subject of community detection.

Meanwhile, for dynamic graphs, it may be of interest to study the evolution of communities over time, once these are identified. Methods in this regard tend to treat the issues of community *detection* and community *tracking* separately, first identifying communities in each time snapshot, which then are inter-linked over time, using temporal smoothing factors [6,18].

[1] Assortativity is a preference for network's nodes to attach to others that are similar or different in some way.

Recognizing the importance of *higher-order* graph metrics, we have included the *Girvan-Newman* algorithm – also known as *edge betweenness* –[19] and *Cluster Percolation Method* (CPM)[20] which can be used to analyse simulation results of both new and existing social models. *Edge betweenness*, born out of the social sciences, works by iteratively removing *central* edges, in order to disconnect the network graph into individual (i.e non-overlapping) communities, while CPM connects adjacent fully connected *k-sized* cliques[2] to discover overlapping communities of nodes.

Given that we have full control of sampling frequency of the network of agent ties during execution, the community *tracking* algorithm *CommTracker* [6] is used. This algorithm matches communities from consecutive snapshots by tracking those nodes it identifies as central to each community, in order to weave together a story of each community's evolution over the course of the snapshots captured.

Yin introduces two important high-order *CommTracker*-reliant network metrics:

- Community **growth** is defined as the correlation between *size* and *age* of a community.
- Community **metabolism** is defined as the correlation between *member stability* and *lifespan* of a community

Community *size* is defined as the number of nodes that are said to belong to it, while *age* is the number of consecutive snapshots a community has been detected in. The *lifespan* of a community is the total number of communities detected in posterior snapshots that can trace their origin to it. Finally, *member stability* (MS) is an index referring to how much a community changes over time calculated as Eq. 1

$$MS(C^t) = \frac{C^t \cap (C_1^{t+1} \cup C_2^{t+1} \cup C_3^{t+1} \ldots \cup C_n^{t+1})}{C^t \cup (C_1^{t+1} \cup C_2^{t+1} \cup C_3^{t+1} \ldots \cup C_n^{t+1})} \quad (1)$$

Where C^t is the set of nodes in a community at time t and C_x^t is the set of nodes in the community x at time t. Finally $C_1^{t+1} \cup C_2^{t+1} \cup C_3^{t+1} \ldots \cup C_n^{t+1}$ is the union of all communities in the posterior time step that can trace their origins to C^t.

These two community-based correlations – *growth* and *metabolism* – are used to successfully differentiate examples of real complex networks as social and non-social. For the social data used in the study (telephone call network, Enron e-mail network, movie actor and co-authorship networks) *growth* was found to be positive, and the *metabolism* index negative. For the non-social datasets used (internet link network, vocabulary network and class invocation relationships from the Apache Ant, Tomcat4 and Tomcat5 projects) this time *growth* was negative while the *metabolism* index positive.

The implications that social and non-social communities have a behavioural signature that can be detected are of great importance to the development of

[2] In graph theory, a clique is a subset of vertices in a graph that is fully connected, that is every two vertices in the subset is connected by an edge.

new agent-based social models. Thus, *CommTracker* is an important addition to an ABSS system, as it helps understand the network dynamics at play in each model and compare them to real-world social network dynamics.

5 Description of the Social Model

In this article we present findings from experiments done on an agent-based model selected from the literature. The goal of these experiments is to assess the effort of implementing new agent-based model and to study the evolution of communities formed from the resulting interaction graph.

The model selected is called *Christakis-Fowler.* This is a parameterized social network model that considers both individual personality traits and peer influence when forming social ties. It was developed to explain the influence alters can have on egos, by studying trend dispersion of obesity and smoking in longitudinal datasets by Christakis and Fowler [21,22].

The model takes into account *homophily* (β_1) – the tendency of egos to create ties to like-alters– and the strength of existing ties to be kept over time (β_0). Algorithm 1 gives a more formal description of the model.

1. A normally distributed trait Y is randomly generated at time t_0 for a population of n agents, where $Y_{t_0} \sim N(50, 100)$
2. Upon agent interaction, differences in Y are computed between agents i, j, where $d_{i,j} = - \mid Y_{i,t_0} - Y_{j,y_0} \mid$.
3. A tie $A_{i,j}$ is created as a function of $d_{i,j}$ using a model based on a latent variable $A^*_{i,j}$. In our implementation of the model, ties are undirected, that is $A_{i,j}$ implies $A_{j,i}$.

$$A_{i,j} = \begin{cases} 1 & \text{if } A^*_{i,j} > \epsilon_{i,j} N(0,1) \\ 0 & \text{otherwise} \end{cases} \tag{2}$$

   Where $A^*_{i,j} = \beta_0 + \beta_1 * d_{i,j}$.
4. Values of Y for each ego are updated as a weighted average of their current value of Y_{t_0}+u and the average value Y_{t_0}+u of their alters, where $u \sim N(0, 5)$.

$$Y_{i,t_{s+1}} = (1-b)(Y_{i,S_s} + rand(0,1)) + b * \frac{\sum_j A_{i,j}(Y_{j,s} + rand(0,1))}{\sum A_{i,j}} \tag{3}$$

   The parameter b is a measure of influence of alters on egos, and the sub-indices i, j and t, indicate the personality traits of the agents involved –ego and alter respectively– at a given time t.
5. All agents update their friendship ties as in step 3.

For a more in-depth look at both model and the impact each change in parameters has on the simulations can be found in [23].

In algorithm 1, the *Update*(Y_i,t_s) function – line 16 – is responsible for giving a new value to the attribute Y, which represents personality traits that egos use as a distance metric to alters. Function 1 is used to implement the *Update* method.

Algorithm 1. Christakis-Fowler Model

```
Parameter: Population of agents 𝒫 each with attributes.
           β0 ←TIE_RETAINING_COEFF
           β1 ←HOMOPHILY_COEFF
           Y_{t_s} ← rand(50,100). Number of simulation steps s_max
1  M ← matrix[𝒫][𝒫]
2  s ← 0
3  for s < s_max do
4      foreach ag_i, ag_j ∈ 𝒫 ∧ ag_i ≠ ag_j do
5          d_{i,j} ← −|Y_{i,t_s} − Y_{j,t_s}|
6          A*_{i,j} ← β0 + β1 d_{i,j}
7          if A*_{i,j} > rand(0,1) then
8              A_{i,j} ← 1
9          end
10         else
11             A_{i,j} ← 0
12         end
13         M[i][j] ← A_{i,j}
14     end
15     for ag_i ∈ 𝒫 do
16         Y_{i,t_s+1} ← Update(Y_i, t_s)
17     end
18 end
```

6 Experimental Results

The system was implemented with parameters taken from the experiments performed in [23], in order to use the result to validate the implementation: $\beta_0 = 0.5$, $\beta_1 = 0.0125$, $b = 0.2$. Once validated, 10 simulations were executed using 1000 agents for 300 steps.

Figure 1 displays a community age versus size correlation graph taken from a sample run. The red line corresponds to the correlation of the data, as can be seen, this correlation is positive. Two parameters play an important role for this behaviour to occur: the tie retaining coefficient β_0 and the alter influence coefficient b. Because the weight difference between β_0 and β_1 favoured the retaining of relationships over time, communities that formed had a tendency to persist over time. The low influence coefficient, on the other had, allowed agents to more freely engage in relationships, adding new agents to the community. As a result, the communities formed are large (b) and persistent (β_0) over time.

Figure 2, contrasts the stability of communities formed, with the trace span of each one. Again, the red line represents the correlation (which is negative). The result is clear: stable communities do not tend to diverge and form new communities. As explained earlier, *CommTracker* matches communities detected in sequential time steps if core nodes of the former are found in the later, or if core nodes of the later were once members of the former. Here too is β_0 the driving parameter as agents do not tend to leave communities –core or otherwise– so

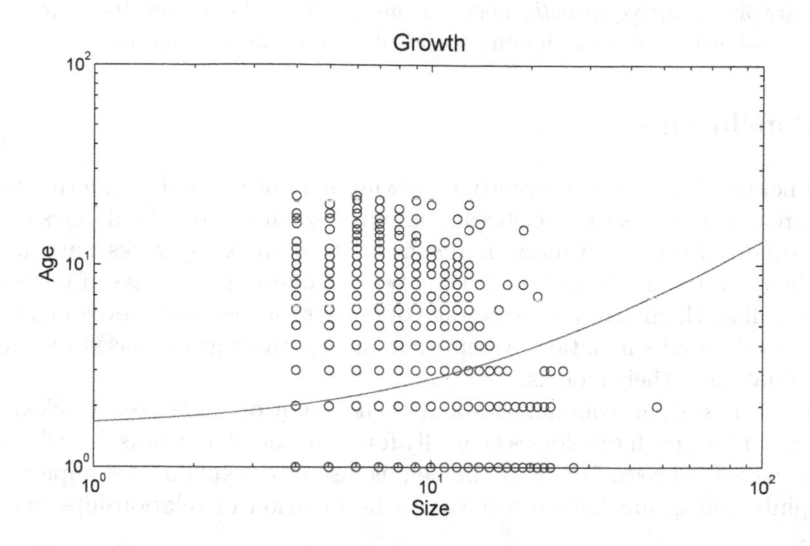

Fig. 1. Sample growth index for communities detected in the Christakis-Fowler social model. Growth= 0.1173.

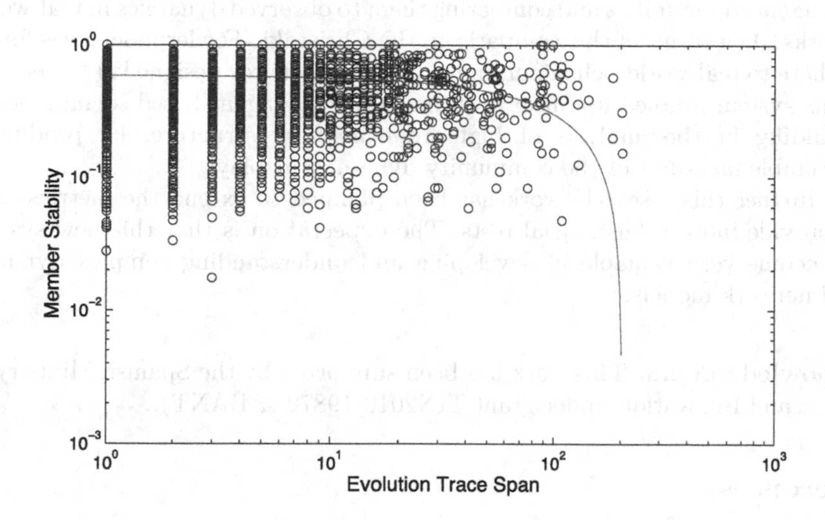

Fig. 2. Sample metabolism index for the same simulation execution as above. Metabolism= −0.1047.

divergence of communities is a rare case. This result may have also been observed had $b > 0.5$, as it would have prevented core nodes from leaving their communities.

Experiments where the growth correlation is positive and metabolism correlation, negative, determines social behaviours. This behaviour, which was detected in 9 out of 10 simulations, matches that observed by Yi [6] in real-world social graphs: positive *growth*, negative *metabolism*. These results confirm the Christakis-Fowler model as having social-like community dynamics.

7 Conclusions

Social network analysis is currently a growing field of research attracting talent from areas like the social sciences, computer science, statistical physics and many others. Recent advances in the field of complex networks have helped to understand the mechanics of social networks, offering new ways of measuring and modelling them. To take advantage of these tools, we have developed a new agent-based social simulation system that incorporates graph-based metrics to the evaluation of their models.

Using this system, communities and their evolution have been tracked over time to test for graph topologies typically found in social networks. The designed model, called *Christakis-Fowler* model, is used to explain the impact that homophily and group influence have on the creation of relationships between agents.

This system was simulated using a new social simulation platform, where social network could be persisted as a sequence of snapshots taken over the duration of the simulation. The model was then analysed by studying several characteristics of the agent communities and comparing them to observed dynamics in real-world networks. As a result of this comparison, the Christakis-Fowler model was found to adhere to real-world behaviour and thus, the simulator designed is correct.

This system proofed useful in both implementing agent-based social models, and aiding in the analysis of high-order network structure, by producing quantifiable measures of the community dynamics at play.

To further this research, work has been planned to extend the metrics used and provide more robust social tests. The expectation is that this new system will become very valuable in developing and understanding complex dynamic social network models.

Acknowledgments. This work has been supported by the Spanish Ministry of Science and Innovation under grant TIN2010-19872 (ABANT).

References

1. Gjoka, M., Kurant, M., Butts, C.T., Markopoulou, A.: Walking in facebook: A case study of unbiased sampling of osns. In: Proceedings of IEEE INFOCOM 2010, pp. 1–9 (March 2010)
2. Klimt, B., Yang, Y.: Introducing the enron corpus. Machine Learning (2004)
3. Abrol, S., Khan, L.: Tweethood: Agglomerative clustering on fuzzy k-closest friends with variable depth for location mining. In: 2010 IEEE Second International Conference on Social Computing (SocialCom), pp. 153–160 (2010)
4. Jahnke, I.: Dynamics of social roles in a knowledge management community. Comput. Hum. Behav. 26(4), 533–546 (2010)

5. Barabasi, A., Jeong, H., Neda, Z., Ravasz, E., Schubert, A., Vicsek, T.: Evolution of the social network of scientific collaborations. Physica A: Statistical Mechanics and its Applications 311(3-4), 590–614 (2002)
6. Wan, Y., Bin, W., Shengqi, Y.: Commtracker: A core-based algorithm of tracking community evolution. Journal of Frontiers of Computer Science and Technology 3(3), 282–292 (2009)
7. Schelling, T.C.: Dynamic models of segregation. Journal of Mathematical Sociology 1, 143–186 (1971)
8. Epstein, J.M., Axtell, R.: Growing artificial societies: social science from the bottom up. The Brookings Institution, Washington, DC (1996)
9. Jurasovic, K., Jezic, G., Kusek, M.: A performance analysis of multi-agent systems. ITSSA 1(4), 335–342 (2006)
10. Li, X., Mao, W., Zeng, D., Wang, F.-Y.: Agent-based social simulation and modeling in social computing. In: Yang, C.C., et al. (eds.) ISI 2008 Workshops. LNCS, vol. 5075, pp. 401–412. Springer, Heidelberg (2008)
11. Davidsson, P.: Agent based social simulation: a computer science view. The Journal of Artificial Societies and Social Simulation 5(1) (January 2002)
12. Snijders, T., van de Bunt, G., Steglich, C.: Introduction to stochastic actor-based models for network dynamics. Social Networks 32(1), 44–60 (2010)
13. Costa, L.D.F., Rodrigues, F.A., Travieso, G., Boas, P.R.V.: Characterization of complex networks: A survey of measurements. Advances in Physics 56(1), 78 (2005)
14. Palla, G., Barabasi, A.L., Vicsek, T.: Quantifying social group evolution. Nature 446(7136), 664–667 (2007)
15. Barabási, A.L., Albert, R.: Emergence of scaling in random networks. Science 286(5439), 509–512 (1999)
16. Caldarelli, G., Capocci, A., De Los_rios, P., Muñoz, M.A.: Scale free networks from varying vertex intrinsic fitness. Physical Review Letters 89 (2002)
17. Fortunato, S.: Community detection in graphs. Physics Reports 486(3-5), 75–174 (2010)
18. Watts, D.J., Strogatz, S.H.: Collective dynamics of 'small-world' networks. Nature 393(6684), 440–442 (1998)
19. Freeman, L.: A set of measures of centrality based on betweenness. Sociometry 40(1), 35–41 (1977)
20. Palla, G., Barabasi, A.L., Vicsek, T.: Quantifying social group evolution. Nature 446(7136), 660–667 (2007)
21. Christakis, N.A., Fowler, J.H.: The collective dynamics of smoking in a large social network. N. Engl. J. Med. 358(21), 2249–2258 (2008)
22. Christakis, N.A., Fowler, J.H.: The spread of obesity in a large social network over 32 years. The New England Journal of Medicine 357(4), 370–379 (2007); Access to full text is subject to the publisher's access restrictions
23. Noel, H., Nyhan, B.: The 'unfriending' problem: The consequences of homophily in friendship retention for causal estimates of social influence. Social Networks 33(3), 211–218 (2011)

Agents in the Browser – Using Agent Oriented Programming for Client Side Development

Alex Muscar

University of Craiova, Blvd. Decebal, nr. 107, RO-200440, Craiova, Romania
amuscar@software.ucv.ro

Abstract. Agent Oriented Programming is in need of an application domain that can give it exposure to a broader audience. Web technologies have matured over the past decade, having reached a point where it is feasible to develop complex, large client side applications, i.e. applications that run in the browser. Together with mobile applications, web applications is a field in which Agent Oriented Programming can shine.

In this paper we propose a dialect of Jason (a subset of Jason extended with promises for asynchronous computations) for client side development. We will give the design rationale of the language and some details on its implementation.

Keywords: agent-oriented programming, asynchrony, promises, programming languages, web development, client side programming.

1 Introduction

The Agent Oriented Programming (AOP) community has seen increased activity in the past couple of years. Research in the area of agent-oriented programming languages has been active, with such examples as [7,14]. Using some of these recent proposals, the agent community has developed real world applications, ranging from web applications [11] to mobile applications [15]. For a couple of reasons we feel that research in AOP languages and their applications has the potential to drive the agent community further as a whole. First, languages are tools, and we are only as skilful as our tools allow us to be. By pushing further what AOP languages can do, we allow the community to tackle ever more challenging problems. Second, by exploring applications outside the conventional areas of agent development, agents as a paradigm might get more traction if they provide better solutions than other existing approaches.

In the past decade web applications have witnessed an unprecedented growth, both in complexity and in ubiquity. As such, the need for better tools is more stringent than ever—the Javascript language is starting to show its shortcomings. Ajax [6], the *de facto* mechanism for achieving asynchrony in web pages requires the use of callbacks to handle action completion. This has given rise to the already infamous "callback hell" problem[1]. The myriad of programming languages targeting the browser is a proof that this niche is still open[2]. We believe that the AOP paradigm is a viable solution to

[1] http://callbackhell.com/

[2] See http://altjs.org/ for a comprehensive list.

A. Bădică et al. (eds.), *Recent Developments in Computational Collective Intelligence*, Studies in Computational Intelligence 513,
DOI: 10.1007/978-3-319-01787-7_8, © Springer International Publishing Switzerland 2014

developing client side applications thanks to the granular and inherently communicative nature of agents.

In this paper we propose a new agent-oriented programming language, which is a dialect of Jason targeting the web browser as runtime. The choice of basing our language on Jason is motivated by the fact that a good deal of the recent efforts in the agent community have been based on the Jason language [1], thus establishing it is a viable choice. By adding support for naturally expressing asynchronous computation in agent languages a whole new area of applications would open up for AOP, from scientific applications like massively parallel simulations to financial applications like *high frequency trading* [16]. There have been some recent efforts to address these issues such as the one proposed by Ricci et al. in [14], but there is still place for other approaches. An early and interesting example of an agent oriented programming language featuring concurrent computation is Go! [3], which has unfortunately mostly gone unnoticed by the community.

The solution we propose avoids the transformation of the program into explicit *continuation passing style* (CPS) [17] inherent when using promises, thus avoiding the "callback hell". We also take into account some of the drawbacks of our proposed approach and investigate some possible solutions.

At the moment of writing this paper, we have a working prototype for the compiler translating the proposed Jason dialect to Javascript. The compiler is implemented in the OCaml programming language, and the source code is available from `https://github.com/muscar/blueprint`.

This paper is structured as follows: in sec. 2 we briefly describe the subset of the Jason language relevant to our dialect and the translation scheme used by the compiler, in sec. 3 we illustrate the main translation strategies of our compiler by using a simple example application, in sec. 4 we motivate the need of a better way of handling asynchronous computations, in sec. 5 we shed some light on the way our dialect handles asynchronous computations, and, finally, in sec. 6 we conclude our paper.

2 Source Language and Translation Scheme

In this section we are going to give more details on the syntax of the source language—the dialect of Jason we are proposing in this paper—and the translation to the target language—Javascript.

Just like Jason, our language has the notions of plans, beliefs, goals, and intentions. A program is made out of three sections:

Beliefs are very similar to facts in Prolog [4] with `&` replacing `,` in conjunctions. The belief base can be manipulated with the aid of the `+` and `-` operators, in the *body* of the plan. They add, and delete respectively, a belief from the belief base. Since these operations usually come in pairs, a deletion followed by an addition, the shortcut operator `-+` is used for this purpose;

Goals are introduced by the `!` operator[3];

[3] To be precise the `!` introduces *achievement* goals. Jason also has *test* goals introduced by the `?` operator, but they are not yet supported by our dialect yet, and as such relevant for the purpose of this paper.

Plans which follow the generic form `triggering_event : context <- body.`, are intended to handle goals. *Triggering events* match goals and message. The only relevant triggering events for this paper are goal addition and message arrival, denoted by atoms with the `+!` and the `+` prefixes.

Internal actions allow agents to call methods from the underlying platform. They are qualified by their package name (like in Java, e.g. `package.action`). Predefined actions do not have a package name, but they retain the leading period (e.g. `.send`). Since the proposed language is a dialect of Jason, we direct the interested readers to [2] for further details.

Figure 1 shows the abstract syntax of the source language[4]. We use *id* to range over atom and functor identifiers, *qid* to range over qualified identifiers (i.e. identifiers with a package name possibly composed of multiple elements separated by the period character), *X* to range over variable identifiers, *a* to range over plan actions, and *s* to range over statements. We also use the standard φ for logical formulas.

Clause c	::=	$+!id(e_0 \ldots e_n) : \varphi \leftarrow s.$	Plan addition
		$-!id(e_0 \ldots e_n) : \varphi \leftarrow s.$	Plan failure
Bel b	::=	$id(e_0 \ldots e_n).$	Belief
Stmt s	::=	a	Plan action
		$do\ a;\ s$	Monadic action
		$s_1; s_2$	Statement sequence
Action a	::=	ϵ	Empty action
		$!id(e_0 \ldots e_n)$	Goal invocation
		$?id(e_0 \ldots e_n)$	Belief query
		$+id(e_0 \ldots e_n)$	Belief update
		$-id(e_0 \ldots e_n)$	Belief deletion
		$qid(e_0 \ldots e_n, X)$	Internal action invocation
Logic φ	::=	$e_1\ op\ e_2$	Binary operator
		$id(e_0 \ldots e_n)$	Belief
		$not\ \varphi$	Negation
		$\varphi_1\ \&\ \varphi_2$	Conjunction
Exp e	::=	n	Numeral
		$true$	Boolean literal
		$false$	Boolean literal
		$"s"$	Literal
		id	Atom
		X	Variable
		$e_1\ op\ e_2$	Binary operator
		$id(e_0 \ldots e_n)$	Structure

Fig. 1. Abstract syntax of the source language

The most important departures from the Jason are the addition of *monadic plan actions*—introduced by the *do* keyword—the lack of Jason's rule language and strong

[4] Lists are not included in Figure 1 since, internally, they are represented as structures of artity 2, with the functor *cons* and the *nil* atom denoting the empty list.

negation. While useful in writing plan contexts, the rule language is orthogonal to plans and goals, thus not essential [2, p. 38-39]. We also introduce the *empty plan action*, ϵ, only as a convenience for translating monadic actions (see Figure 2). Also, the last argument of an internal action invocation must be a variable, which will be bound to the result of the call to the native method (see Figure 2). The compiler enforces this rule.

Figure 2 shows the most relevant rules for translating the source language introduced in Figure 1 to Javascript.

$$
\begin{aligned}
[\![+!id(e_0 \ldots e_n) : \varphi \leftarrow s.]\!] &= \textbf{function } id(p_0 \ldots p_n) \, \{ \\
&\qquad \textbf{if } (\textbf{unify}(p_0, [\![e_0]\!]) \; \&\& \; \ldots \; \&\& \; \textbf{unify}(p_n, [\![e_n]\!]) \; \&\& \; [\![\varphi]\!]) \, \{[\![s]\!]\} \\
&\quad \} \\
[\![id(e_0 \ldots e_n).]\!] &= \textbf{var } id = \textbf{new term}(\textbf{tag}(id), \textbf{name}(id), [\![e_0]\!], \ldots, [\![e_n]\!]) \\
[\![!id(e_0 \ldots e_n)]\!] &= id([\![e_0]\!], \ldots, [\![e_n]\!]) \\
[\![?id(e_0 \ldots e_n)]\!] &= \textbf{unify}(\textbf{item\_0}(id), [\![e_0]\!]); \ldots; \textbf{unify}(\textbf{item\_n}(id), [\![e_n]\!]) \\
[\![+id(e_0 \ldots e_n)]\!] &= id = \textbf{new term}(\textbf{tag}(id), \textbf{name}(id), [\![e_0]\!], \ldots, [\![e_n]\!]) \\
[\![+id(e_0 \ldots e_n)]\!] &= id = \textbf{null} \\
[\![qid(e_0 \ldots e_{n,)}]\!] &= id([\![e_0]\!], \ldots, [\![e_n]\!]) \\
[\![qid(e_0 \ldots e_n, X)]\!] &= \textbf{unify}(X, qid([\![e_0]\!], \ldots, [\![e_n]\!])) \\
[\![do\ a;\ \epsilon]\!] &= \textbf{bind}_{ctx}([\![a]\!], \textbf{function} \, () \, \{\}) \\
[\![do\ a;\ s]\!] &= \textbf{bind}_{ctx}([\![a]\!], \textbf{function} \, () \, \{[\![s]\!]\}) \\
[\![s_1; s_2]\!] &= [\![s_1]\!]; [\![s_2]\!] \\
[\![id(e_0 \ldots e_n)]\!] &= \textbf{new term}(\textbf{tag}(id), \textbf{str}(id), [\![e_0]\!], \ldots, [\![e_n]\!]) \\
[\![X]\!] &= \textbf{new variable}(\textbf{name}(X))
\end{aligned}
$$

Fig. 2. Selected rules for translating the source language to Javascript

The most interesting ones are related to monadic actions and internal actions. Monadic actions are translated as a call to the appropriate $\textbf{unify}_{ctx}$ method of the monadic context of the current plan (see sec. 5). Because monads naturally introduce sequencing, a monadic action has to be followed by another plan action. To handle the case in which a monadic action is the last action in a clause, we introduce the empty pan action, ϵ, so that, conceptually, it is followed by an action whose translation to Javascript does nothing.

3 Use Case

This section walks through the implementation of a short web application using the proposed language. The application used the Facebook Javascript SDK[5] to fetch a list of status messages for the visitor and the jQuery[6] awesomeCloud plugin[7] to create a word cloud[8] with the most frequent words used by the user in her statuses. We will

[5] `https://developers.facebook.com/docs/reference/javascript/`

[6] `http://jquery.com/`

[7] `http://indyarmy.com/awesomeCloud/`

[8] A word cloud is a visual representation for text data, typically used to depict keyword metadata (tags) on websites, or to visualise free form text. Tags are usually single words, and the importance of each tag is shown with font size or colour[9].

use this example to illustrate the translation process described in sec. 2 as well as to give the reader a feel of how our language integrates with Javascript libraries and how it fares for developing client side applications. A running example of can be found at `http://muscar.github.io/blueprint/test.html`[9].

3.1 Translating Agents

Agents are translated as Javascript constructor functions [5]. One input file corresponds to one Javascript constructor function. The agent name (hence the constructor function's name) is derived from the source file's name, e.g. in our case the source file is named `word_cloud.b`, thus the agent name is `word_cloud` (see Figure 3).

```
function word_cloud() {
    var $self = this;

    // Agent beliefs and plans
}
```

Fig. 3. The agent constructor function

In order to avoid Javascript's infamous scoping problems [5], the compiler generates a `$self` field so that it can refer to the agent object inside plans.

3.2 Translating Plans

As mentioned before, a word cloud needs the weights, in our case the frequency, of the words it is going to display. As such we need to write plans to split each status message into words and count each apparition of an individual word. We will defer the plan to actually fetch the status messages until we talk about asynchrony.

Figure 4 shows the plans to compute word frequencies in the user's status messages. The plan `+!collect` iterates over a list of status messages, and creates a single list of all the words in all the status messages. Note that the plan allows duplicates. it is the job of the `+!count_words` plan to count number of apparitions of each word in the list produced by `+!collect`.

The code in Figure 4 should be very familiar to Jason users. The most interesting aspect are the calls to the various builtin actions defined by the language's standard[10] library (in the package `blueprint.lang`) as well as to the function `window.makeCloud` defined in Javascript. This shows that the application developer can interact with the

[9] Please note that due to some incompatibilities the application might not work on Firefox. It works however in recent versions of Google Chrome, Apple Safari and Microsoft Internet Explorer.

[10] The language comes with a standard library for runtime operations. At the time of writing this paper the standard library is limited to the functionality necessary to implement the proof of concept example described later in this paper.

```
+!collect([], Acc, Acc).
+!collect([Status|Statuses], Acc, R)
<- blueprint.lang.str.split(Status, Words);
   blueprint.lang.list.concat(Acc, Words, Acc1);
   !collect(Statuses, Acc1, R).

+!count([], D)
<- window.makeCloud("#cloud", D, _).
+!count([W|Ws], D)
<- blueprint.lang.dict.find_default(D, W, 0, C);
   blueprint.lang.dict.add(D, W, C + 1, _);
   !count(Ws, D).

+!count_words(Ws)
<- blueprint.lang.dict.make(D);
   !count(Ws, D).
```

Fig. 4. Plans to compute word frequencies in the user's status messages

host platform both by defining internal actions which act as wrappers for complex functionality defined in the host language, in a way similar to Jason, as well as by directly calling functions defined in the host language, which is not possible in Jason. This allows for simplified interoperability with the host platform. As mentioned in sec. 2, native calls must be provided an extra parameter, a variable, to be bound to the result of the call. If the developer wishes to ignore the result she can pass a wildcard variable, _.

The compiler translates plans as Javascript object methods. Since the generated code follows the same pattern, and to conserve space, we will only show the translation of the `+!collect` plan to Javascript in Figure 5.

Plan clauses are compiled as nested conditional tests. Variables appearing in the triggering events, in the context as well as in the body of clause are collected and initialised to objects of the class `$variable`, which is defined in the standard library. The triggering event's parameters are compiled as calls to the `$unify` library method which implements the classic unification algorithm. The `$unify` method returns `true` if the unification was successful, `false` otherwise. This allows it to be used as a conditional test. The compiler also translates the context condition to logic expressions and uses them together with the calls to `$unify` generated for the plan arguments in order to decide if the plan body is applicable. In our example, since the clause has an empty context, the condition generated for it is just `true`. Because the unification process destructively updates variables and because all the clauses share the same variables, when a new plan clause needs to be tested, the variables need to potentially be unbound. This is achieved by calling the `unbind` method on the variable object.

Again, as mentioned in the corresponding translation rule in Figure 2, action invocations are compiled as calls to the `$unify` library function:

```
$unify(blueprint.lang.str.split(Status.getValue()), Words);
```

```
this.collect = function (param0, param1, param2) {
    var Acc = new $variable('Acc');
    var Acc1 = new $variable('Acc1');
    var R = new $variable('R');
    var Status = new $variable('Status');
    var Statuses = new $variable('Statuses');
    var Words = new $variable('Words');
    Acc.unbind();
    Acc1.unbind();
    R.unbind();
    Status.unbind();
    Statuses.unbind();
    Words.unbind();
    if ($unify(param0, $nil) &&
        $unify(param1, Acc) &&
        $unify(param2, Acc) &&
        true) {
        } else {
            Acc.unbind();
            Acc1.unbind();
            R.unbind();
            Status.unbind();
            Statuses.unbind();
            Words.unbind();
        if ($unify(param0, new $cons(Status, Statuses)) &&
            $unify(param1, Acc) && $unify(param2, R) &&
            true) {
            $unify(blueprint.lang.str.split(Status.getValue()), Words);
            $unify(blueprint.lang.list.concat(Acc.getValue(),
                                            Words.getValue()), Acc1);
            $self.collect(Statuses, Acc1, R)
        } else {
            $error("plan collect failed")
        }
    }
};
```

Fig. 5. Plan translation to Javascript

The `getValue` method returns the value of a variable if it is bound, or throws an exception otherwise.

Finally, if no clause of the plan is applicable, and there are no failure clauses, the compiler will generate code to signal an error that the current plan has failed.

4 Dealing with Asynchrony

When it comes to asynchrony, Ajax is the standard method in web applications. By using support provided by the browser, web applications are able to asynchronously send and receive data without blocking the page. In order to be notified when the request is made, applications need to register callbacks for the requests. This leads to a style of programming that makes composing multiple asynchronous computations difficult.

In this section we are going to look at our concrete example, asynchronously fetching the user's statuses from Facebook, and analyse the approach to see if it can be generalised. Unlike the previous section, we will start from a Javascript solution and only then look at the agent version of the code (see Figure 6).

```
function makeCloud() {
    FB.api('/me/statuses', function (result) {
        // process the statuses
    });
}
```

Fig. 6. Plan translation to Javascript

In Figure 6 we use the standard Ajax pattern: we call an asynchronous method, `FB.api`, and pass it a callback which will be invoked when the call is done. In the callback we can continue our work. This forces the developer to explicitly write the code in continuation passing style. While in our particular example this does not seem like a major issue let us extend our example to also fetch the user likes from Facebook (see Figure 7).

```
function makeCloud() {
    FB.api('/me/statuses', function (result1) {
        FB.api('/me/likes', function (result2) {
            // process the statuses
            // process the likes
        });
    });
}
```

Fig. 7. Plan translation to Javascript

There are two problems with the approach in Figure 7: first, the code can easily get out of hand if we have multiple asynchronous calls or more complex logic, and, second, the second call to the Facebook API will not get executed until the first one returns, which is a waste of time. One way to solve this is the jQuery Deferred[11] library.

Figure 8 illustrates the Deferred approach for the implementation in Javascript. While this accomplishes our goals, programming in a linear style and efficiently fetching both resources concurrently, the code has doubled in size, adding a lot of syntactic noise.

[11] `http://api.jquery.com/category/deferred-object/`

```
function makeCloud() {
    var statusesP = jQuery.Deferred();
    var likesP = jQuery.Deferred();

    FB.api('/me/statuses', function (result) {
        statusesP.resolve(result);
    });

    FB.api('/me/likes', function (result) {
        likesP.resolve(result);
    });

    jQuery.when(statusesP, likesP).then(function (statuses, likes) {
        // process the statuses
        // process the likes
    });
}
```

Fig. 8. Plan translation to Javascript

5 Implicit Promises

This work builds on our previous efforts of extending Jason with promises [12]. As such, we will not enter into great details on the subject of promises. We direct the interested reader to that paper for further details. While most implementations of promises follow some common principles, there is no agreed-upon definition of what a promise is. Using monads to give structure to promises seems promising in the light of projects such the *computation expressions* in F# [18], the C# asynchronous model [10], and Scala promises [8]. Before we go on any further we must first briefly introduce monads.

Monads are structures that represent computations. They are usually defined as triples composed of a *type constructor* with two associated operations called *return* and *bind*.

Having promises form a monad opens up some interesting options for both the syntax—F#'s computational expressions or Haskell's *do notation* [13]— and semantics —monads can be easily composed in a series of interesting ways—of our dialect. Next we will further explore the implications on the design of our solution.

Monads have been called "programmable semicolons" because of the syntactic sugar that they have associated in Haskell, the *do notation* [13]. The do notation allows Haskell developers to define custom semantics on a per-function basis while maintaining a standard notation.

We will employ a similar scheme in our dialect, by specifying the context in which a plan is going to be executed via annotations. This will allow the programmer to write simpler and shorter code, while allowing the compiler to rewrite the code in order to do the appropriate plumbing. This approach is meant to address the inherent *inversion of control*[12] that arises in concurrent scenarios (see sec. 2).

[12] Linear control flow is replaced by a scheme where each function call (or predicate/plan) receives an additional functional argument, its "continuation", witch is called instead of normally returning a value.

Figure 9 illustrates our approach. Returning to our initial example of fetching only the user statuses, the plan make_cloud is annotated as executing in the promise context. This changes the semantics of the actions inside the plan, making them behave as promise computations instead of regular Jason code.

```
+!make_cloud[context(promise)]
<- do facebook.statuses(Statuses);
   !collect(Statuses, [], Words);
   !count_words(Words).
```

Fig. 9. A plan in the promise monad

As specified by the rewrite rules in sec. 2, the compiler rewrites the code in Figure 9 to the Javascript code in Figure 10. The call to promise.bind method is the most interesting part of the translation. Because the plan has the context(promise) annotation, the compiler will treat the monadic actions in the plan body, i.e. the ones prefixed by the do keyword, as calls to the bind method of the constructor specified in the context. The constructor is responsible for implementing the desied semantics, and can (and should) be part of a library.

```
this.make_cloud = function () {
  var Statuses = new $variable('Statuses');
  var Words = new $variable('Words');
  Statuses.unbind();
  Words.unbind();
  if (true) {
    promise.bind(facebook.statuses(Statuses), function () {
      $self.collect(Statuses, $nil, Words);
      $self.count_words(Words);
    })
  } else {
    $error("plan make_cloud failed")
  }
};
```

Fig. 10. Rewriting for monadic actions

This approach scales nicely to multiple asynchronous calls as seen in Figure 11. Compared to the 16 lines of code necessary to implement the same in Javascript (see Figure 8), the Jason solution is much terser at only 5 lines of code.

```
+!make_cloud[context(promise)]
<- do facebook.statuses(Statuses);
   do facebook.likes(Likes);
   // process statuses
   // process likes
```

Fig. 11. Multiple asynchronous calls

6 Conclusion and Future Work

In this paper we presented a dialect of Jason that compiles to Javascript together with a non-intrusive extension for asynchronous computations using a monad for structuring promises. Being able to easily compose asynchronous computations offers a big advantage for real world scenarios where agents need to use resource that imply latencies, e.g. web services.

As a future direction of research we would like to extend the monadic model from promises to generic computations. It has been proven by other projects that such an approach might be beneficial for language experimentation.

Another direction is investigating tighter integration of the agent with the host—the browser. One interesting venue is exploring web workers from the upcoming HTML5 standard as a possible back-end for running agents.

References

1. Bordini, R.H., Hübner, J.F., Vieira, R.: Jason and the golden fleece of agent-oriented programming. In: Bordini, R.H., Dastani, M., Dix, J., Fallah-Seghrouchni, A.E. (eds.) Multi-Agent Programming, Multiagent Systems, Artificial Societies, and Simulated Organizations, vol. 15, pp. 3–37. Springer (2005)
2. Bordini, R.H., Wooldridge, M., Hübner, J.F.: Programming Multi-Agent Systems in AgentSpeak using Jason (Wiley Series in Agent Technology). John Wiley & Sons (2007)
3. Clark, K.L., McCabe, F.G.: Go! – a multi-paradigm programming language for implementing multi-threaded agents. Annals of Mathematics and Artificial Intelligence 41(2-4), 171–206 (2004)
4. Clocksin, W.F., Mellish, C.S.: Programming in Prolog, 4th edn. Springer (1994)
5. Crockford, D.: JavaScript: The Good Parts. O'Reilly Media, Inc. (2008)
6. Garrett, J.J.: Ajax: A new approach to web applications (2005), http://www.adaptivepath.com/ideas/ajax-new-approach-web-applications
7. Grigore, C.V., Collier, R.W.: Af-raf: an agent-oriented programming language with algebraic data types. In: Proceedings of the Compilation of the co-located Workshops on DSM 2011, TMC 2011, AGERE! 2011, AOOPES 2011, NEAT 2011, & VMIL 2011, SPLASH 2011 Workshops, pp. 195–200. ACM, New York (2011), http://doi.acm.org/10.1145/2095050.2095081, doi:10.1145/2095050.2095081
8. Haller, P., Prokopec, A., Miller, H., Klang, V., Kuhn, R., Jovanovic, V.: Futures and promises, http://docs.scala-lang.org/overviews/core/futures.html
9. Halvey, M.J., Keane, M.T.: An assessment of tag presentation techniques. In: Proceedings of the 16th International Conference on World Wide Web, WWW 2007, pp. 1313–1314. ACM, New York (2007), http://doi.acm.org/10.1145/1242572.1242826, doi:10.1145/1242572.1242826

10. Microsoft: Asynchronous programming with async and await (2013), http://msdn.microsoft.com/en-us/library/vstudio/hh191443.aspx
11. Minotti, M., Piancastelli, G., Ricci, A.: Agent-oriented programming for client-side concurrent web 2.0 applications. In: Cordeiro, J., Filipe, J. (eds.) WEBIST 2009. LNBIP, vol. 45, pp. 17–29. Springer, Heidelberg (2010)
12. Muscar, A.: Extending jason with promises for concurrent computation. In: Fortino, G., Badica, C., Malgeri, M., Unland, R. (eds.) Intelligent Distributed Computing VI. SCI, vol. 446, pp. 41–50. Springer, Heidelberg (2012), http://dx.doi.org/10.1007/978-3-642-32524-3_7
13. O'Sullivan, B., Goerzen, J., Stewart, D.: Real World Haskell. O'Reilly Media, Inc. (2008)
14. Ricci, A., Santi, A.: Designing a general-purpose programming language based on agent-oriented abstractions: the simpal project. In: Proceedings of the Compilation of the Co-located Workshops on DSM 2011, TMC 2011, AGERE! 2011, AOOPES 2011, NEAT 2011, & VMIL 2011, SPLASH 2011 Workshops, pp. 159–170. ACM, New York (2011), doi:10.1145/2095050.2095078
15. Santi, A., Guidi, M., Ricci, A.: JaCa-android: An agent-based platform for building smart mobile applications. In: Dastani, M., El Fallah Seghrouchni, A., Hübner, J., Leite, J. (eds.) LADS 2010. LNCS, vol. 6822, pp. 95–114. Springer, Heidelberg (2011)
16. International Organization of Securities Commissions, T.C.: Regulatory issues raised by the impact of technological changes on market integrity and efficiency. Tech. rep., International Organization of Securities Commissions (2011), http://www.iosco.org/library/pubdocs/pdf/IOSCOPD354.pdf
17. Sussman, G.J., Steele Jr. G.L.: Scheme: an interpreter for extended lambda calculus. MIT AI Memo 349. Massachusetts Institute of Technology, Cambridge (1975)
18. Syme, D., Granicz, A., Cisternino, A.: Expert F# 2, 1st edn. Apress, Berkely (2010)

Part II
Intelligent Computational Methods

Handwritten Numerical Character Recognition Based on Paraconsistent Artificial Neural Networks

Sheila Souza[1,2,3] and Jair Minoro Abe[2,4]

[1] School of Medicine, University of São Paulo, São Paulo, Brazil
[2] Institute for Advanced Studies, University of São Paulo, São Paulo, Brazil
[3] PRODESP, Data Processing Company of São Paulo State, São Paulo, Brazil
sheinara@gmail.com
[4] Graduate Program in Production Engineering, ICET - Paulista University, São Paulo, Brazil
jairabe@uol.com.br

Abstract. This paper presents an automated computational process able to recognize a handwritten numerical characters and Magnetic Ink Character Recognition used on bank checks based on Paraconsistent Artificial Neural Networks. The methodology employed was chosen for being a tool able to work with imprecise, inconsistent and paracomplete data without trivialization. The recognition process is performed from some character features previously selected based on some Graphology and Graphoscopy techniques and, the analysis of such features as well as the character recognition are performed by Paraconsistent Artificial Neural Networks.

Keywords: Artificial Intelligence, Pattern Recognition, Character Recognition, Handwritten Numerical Character Recognition, Artificial Neural Networks, Paraconsistent Artificial Neural Networks.

1 Introduction

Computer pattern recognition is one of the most important Artificial Intelligence tools in numerous knowledge areas with applications in several themes, including the character recognition. Although there are several studies on character recognition [4][9][15], we have chosen to study this field due to its intrinsic importance and constant improvement, besides enabling adjustments to the recognition of different types of signals.

The character recognition is one of the most known and explored pattern recognition's modalities which consists, roughly speaking, in extracting some features from a character group aiming to try to reproduce human being ability to read texts.

The first applications in Optical Character Recognition (OCR) started in the early 1960's. Such systems were developed to read characters having a pre-determined pattern data (e.g. IBM 1418) and then the methods were improved and new systems were developed to recognize hand-printed characters as IBM 1287, RETINA (by Recognition Equipment Inc) and H8959 (by Hitachi). Later, some commercial

A. Bădică et al. (eds.), *Recent Developments in Computational Collective Intelligence*, Studies in Computational Intelligence 513,
DOI: 10.1007/978-3-319-01787-7_9,

systems have appeared to recognize poor-print-quality characters and hand-printed characters for a large category of character set Kanji OCR (by Toshiba) and CLL-2000 (by Sanyo Electric Co. Ltd). Over the last years the techniques have been improved to develop systems to recognize noisy documents, color document, handwritten characters and complex documents with texts, graphics, tables and mathematical symbols, etc. [6][14]

The character recognition technique classifies the characters through their features. Therefore, due to noises interference, the major difficulty concentrates on determining the features group liable to extraction.

Considering these issues, the performance of a pattern recognition automated system depends fundamentally on the quality of the original and digital documents and also on a system which often face with imprecise, conflicting and paracomplete data. Therefore, we have chosen to use Paraconsistent Artificial Neural Networks because they have been considered as an important analysis tool for applications involving data with such features.[3][11][12][13]

The Paraconsistent Artificial Neural Network (PANNs) is based on Artificial Neural Networks and Paraconsistent Annotated Evidential Logic Eτ [8] which is able to handle concepts such as imprecise, inconsistency and paracomplete in its interior without trivialization [2]. The PANNs is composed by Paraconsistent Artificial Neural Units (PANUs) with different functions such as connection, learning, memorization, etc., which process the input signs of the network coming from as favorable and contrary evidence degrees. These units are known as a cluster of Paraconsistent Artificial Neural Cells interconnected with each other that analyze and model electric signs.[8]

2 Automated Computational Process

The Automated Computational Process proposed consists in a system able to extract some features of interest in evidence degrees format from a scanned image preprocessed and, from these features, recognize the character based on PANNs.

The feature extraction process of the proposed system is performed based on fundamentals of Graphology and Graphoscopy techniques [5][9] and some of their aspect will be used throughout this work.

For the numerical character study with a present pattern was adopted the Magnetic Ink Character Recognition (Fig. 1) used on Brazilian bank checks to codify data from the customer's bank account.

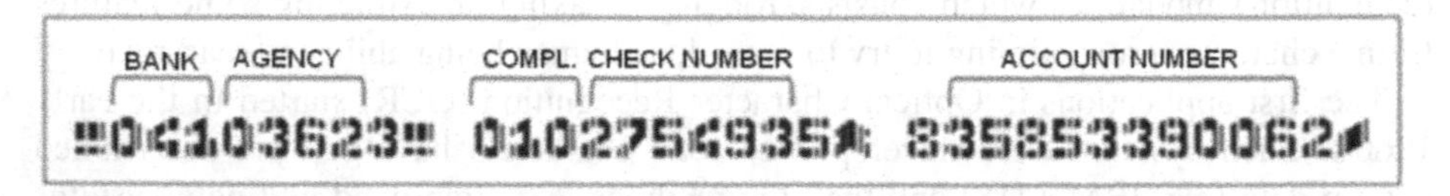

Fig. 1. Magnetic Ink Character Recognition (MICR) example used on Brazilian bank checks

3 Methods

For the proposed system we consider that the recognition process is divided in five steps: 1) Image acquisition, 2) Image preprocessing, 3) Image mapping in evidence degrees, 4) Extracting image features, and 5) Image recognition.

In this paper we focus on the third step ahead. However, as a simple image preprocessing, we apply the average filter among neighbor pixels, image's binarization, the trimming of image's white borders, and image resizing to 38x30 pixels.

The architecture system is presented in Fig. 2. Here we assume a preprocessed image as an input.

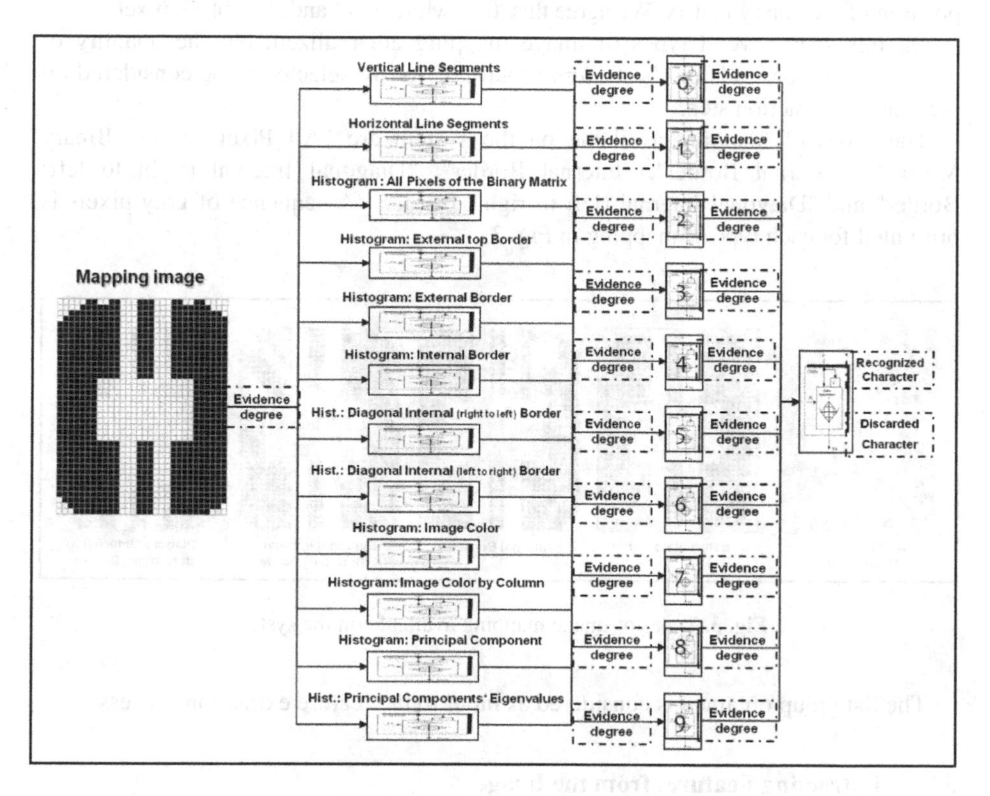

Fig. 2. The System Architecture

The first layer is connected to the third step of the recognition process, where the image is mapped in evidence degrees format. The second layer is connected to the fourth step, where the features are extracted from the image taking into account concepts of Graphology and Graphoscopy techniques. The next two layers are connected to the fifth step. On the third layer, it compares the features of each pattern character to the presented image and calculates an evidence degrees of recognition for each pattern character. On the last layer, it applies a paraconsistent analysis among resulting recognition evidence degree from each pattern character and identifies the recognized and discarded characters.

The following presents the procedures performed at each architecture layer.

3.1 Mapping Image in Evidence Degrees

The mapping image in evidence degrees consists in analyzing the pixels from the binary matrix of the image in several situations and obtaining evidence degree lists which represent an image projection. Each evidence degree is a value belonging to the real interval [0, 1] that represents the presence of a black or white pixel in a determined position of the binary matrix. We agree that '0' is white pixel and '1' is black pixel.

On this step, several types of image mapping are realized, but the quantity of different types depends on the feature group previously selected to be considered on the feature extraction step.

The types of mapping available on the system are "All Pixels of the Binary Matrix", "External Border", "Internal Border", "Diagonal Internal (right to left) Border" and "Diagonal Internal (left to right) Border". A sequence of gray pixels is presented for each type of mapping in Fig. 3.

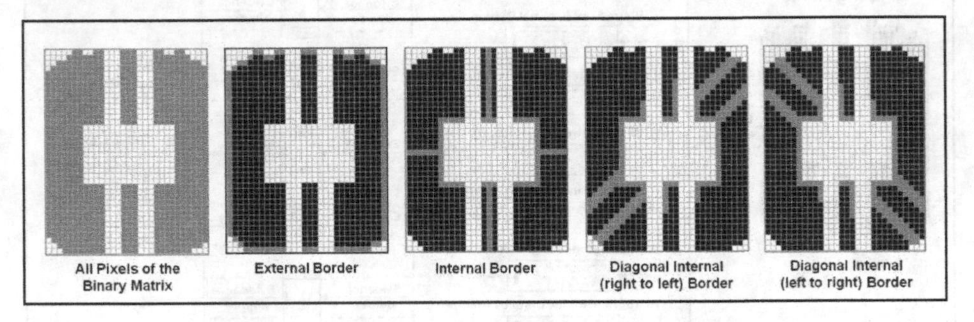

Fig. 3. Types of image mapping available on the system

The list group obtained is considered as input to the feature extraction process.

3.2 Extracting Features from the Image

The extracting features from the image consists in grouping some character features mapped in the third step and turn them into datasets expressed in evidence degrees in a way that they can be transmitted and received by the recognition step.

As in the recognition process, the character features play a fundamental role, we've chosen some of them based on the study of some Graphology and Graphoscopy techniques as "path and curvilinear values" to compose the feature extraction process. The features available on the system to be extracted are "Vertical Line Segments",

"Horizontal Line Segments", "Histogram: All Pixels of the Binary Matrix", "Histogram: External top Border", "Histogram: External Border", "Histogram: Internal Border", "Histogram: Diagonal Internal (right to left) Border", "Histogram: Diagonal Internal (left to right) Border", "Histogram: Image Color", "Histogram: Image Color by Column", " Histogram: Principal Components", "Histogram: Principal Components' Eigenvalues" as presented on the second layer of the architecture system showed in Fig.2.

For the first two features, "Vertical Line Segments" and "Horizontal Line Segments", the system uses PANNs to identify vertical and horizontal line segments and turn them into evidence degrees.

For the others, the evidence degrees groups obtained on mapping image step are neatly grouped to represent some image histograms according to Fig. 4.

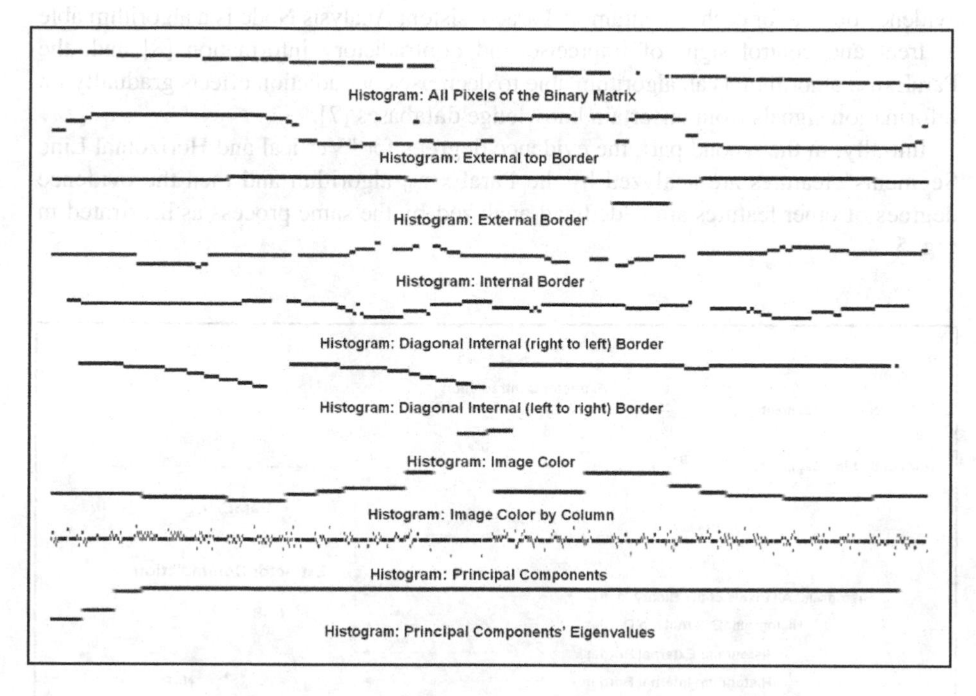

Fig. 4. Image histograms created by the system in fourth step

At the end of this process, we have evidence degree lists to represent each feature extracted from the image and then compose the image recognition step input.

3.3 Image Recognition

On the recognition process, the evidence degrees obtained from the feature extraction step are compared to the evidence degrees of each pattern character, in other words, a paraconsistent processing is realized for each pattern character and its output represents its recognition evidence degree.

The image recognition step is divided in two parts: a) the comparison between pattern character features and presented character features (third layer presented on Fig. 2) and, b) a paraconsistent analysis which consists in calculating a single evidence degree through $\text{ParaExtr}_{\text{ctr}}$ algorithm, that represents pattern character recognition evidence degree (fourth layer presented on Fig. 2).

On the first part, the features considered here depend on the feature group previously selected to be extracted on the image extraction step. For "Vertical Line Segments" and "Horizontal Line Segments" features we use a PANN to compare the features between pattern character and presented character. For the histograms, we use a Paraconsistent Analysis Node to calculate the distance between the pattern character histogram and presented character histogram and, the $\text{ParaExtr}_{\text{ctr}}$ algorithm to calculate a single evidence degree for each histogram. A Paraconsistent Analysis Node is a algorithm able to treat and control signs of imprecise and contradictory information [8] and, the $\text{ParaExtr}_{\text{ctr}}$ algorithm, is an algorithm able to decrease contradiction effects gradually on information signals from uncertain knowledge databases [7].

Initially, in the second part, the evidence degrees for "Vertical and Horizontal Line Segments" features are analyzed by the $\text{ParaExtr}_{\text{ctr}}$ algorithm and then the evidence degrees of other features are added and analyzed by the same process as illustrated in Fig. 5.

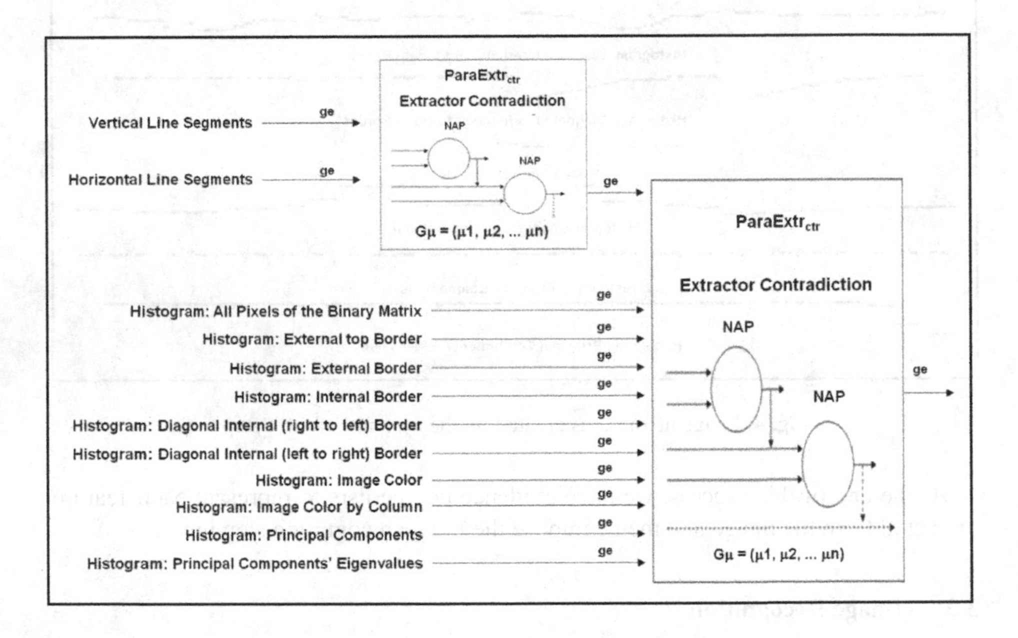

Fig. 5. Paraconsistent analysis architecture

At the end of this process, each pattern character has a single recognition evidence degree. The pattern character with the maximum value represents the recognized character by the system and the pattern character with the minimum value represents the discarded character by the system (fourth layer of the Fig. 2).

4 Results

As the recognition process is performed based on pattern character features, the first step before starting the tests is to create a pattern character feature database by presenting to the network only the pattern character sets (Fig. 6).

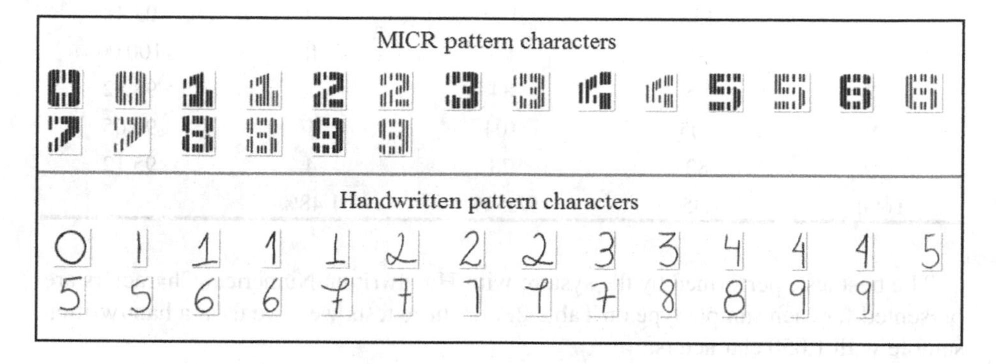

Fig. 6. Pattern character sets

After create the pattern character feature database, some tests were performed considering different features combinations and the system performance was evaluated with real data as Brazilian checks batches and handwritten numerical scanned characters.

The best performance of the system for both types of characters (numerical characters with a present pattern and handwritten numerical scanned characters) was obtained considering only the following features: "Histogram: External Border", "Histogram: Diagonal Internal (right to left) Border" and "Histogram: Diagonal Internal (left to right) Border".

The best performance for characters with a present pattern (MICR characters from scanned original Brazilian bank checks) presented 98.52% hits and 99.7% accuracy and, for handwritten numerical characters (scanned handwritten numerical characters from different writers), presented 91.14% hits and 98.2% accuracy.

The best tests performed by the system with the nine MICR digits are presented for each sample type on Table 1. For these tests we have used a MICR sample with 1,354 characters.

Table 1. Results obtained in tests with the nine MICR digits

Selected features for the test: "Histogram: External Border", "Histogram: Diagonal Internal (right to left) Border" and "Histogram: Diagonal Internal (left to right) Border"				
Sample Type	**Sample Size**	**Hits**	**Errors**	**% Hits**
0	336	336	0	100.00
1	187	185	2	98.93
2	98	96	2	97.96
3	122	121	1	99.18
4	115	115	0	100.00
5	138	129	9	93.48
6	86	86	0	100.00
7	85	84	1	98.82
8	105	104	1	99.05
9	82	78	4	95.12
Total	1,354	98.52%	1.48%	

The best tests performed by the system with Handwritten Numerical Characters are presented for each sample type on Table 2. For these tests we have used a handwritten sample with 1,050 characters.

Table 2. Results obtained in tests with Handwritten Numerical Characters

Selected features for the test: "Histogram: External Border", "Histogram: Diagonal Internal (right to left) Border" and "Histogram: Diagonal Internal (left to right) Border"				
Sample Type	**Sample Size**	**Hits**	**Errors**	**% Hits**
0	109	109	0	100.00
1	92	62	30	67.39
2	121	117	4	96.69
3	134	121	13	90.30
4	85	76	9	89.41
5	103	79	24	76.70
6	104	103	1	99.04
7	88	87	1	98.86
8	116	109	7	93.97
9	98	94	4	95.92
Total	1,050	91.14%	8.86%	

In some previous studies in character recognition, we found out that the VeriFone's MICR reader system - based on neural networks - reaches 99.6% accuracy according to Abdleazeem &El-Sherif [16] and Öksüz [17] presents a handwritten character recognition system with 90.21% correct recognition rate which uses directional and positional information calculated from pen-tip positions by using USB CCD Camera for character classification.

Taking into consideration the industrial point of view in [18] the A2iA 'CheckReader' has recognized with 96% accuracy for constrained handwritten text outperforming in accuracy some of the prominent vendors in the automatic bank check processing area as Mitek, Parascript and SoftPro by using neural networks.

The evidence collected during the tests allows an acceptance of the proposed paraconsistent model. As pointed above, it shows that the obtained model can provide results as good as the other recognition systems.

5 Conclusion

The computational system presented in this work to recognize characters with a present pattern and handwritten numerical characters by using PANNs performed 98.52% hits to characters with a present pattern and 91.14% hits to handwritten numerical characters. We intend to apply such ideas to correlated themes in pattern recognition, so the results obtained up to here encourage us to do so.

References

1. Abe, J.M.: Foundations of Annotated Logics. PhD thesis, University of São Paulo, Brazil (1992) (in Portuguese)
2. Abe, J.M.: Some Aspects of Paraconsistent Systems and Applications. Logique et Analyse 157, 83–96 (1997)
3. Abe, J.M., Lopes, H.F.S., Anghinah, R.: Paraconsistent Artificial Neural Network and Alzheimer Disease: A Preliminary Study. Dementia & Neuropsychologia 3, 241–247 (2007)
4. Abdleazeem, S., El-Sherif, E.: Arabic handwritten digit recognition. International Journal of Document Analysis and Recognition (IJDAR) 11(3), 127–141 (2008)
5. Amend, K., Ruiz, M.S.: Handwriting Analysis: The Complete Basic Book. Franklin Lakes, NJ (1980)
6. Bortolozzi, F., Brittto Jr., A.S., Oliveira, L.E.S., Morita, M.: Recent Advances in Handwriting Recognition. In: Pal, U., Parui, S.K., Chaudhuri, B.B. (Org.) Document Analysis, Chennai, pp. 1–30 (2005)
7. Da Silva Filho, J.I.: Paraconsistent algorithm extractor of contradiction's effects - $\text{ParaExtr}_{\text{ctr}}$. Seleção Documental 15, 21–25 (2009) (in Portuguese)
8. Da Silva Filho, J.I., Torres, G.L., Abe, J.M.: Uncertainty Treatment Using Paraconsistent Logic – Introducing Paraconsistent Artificial Neural Networks. IOS Press, Netherlands (2010)

9. Fujisawa, Y., Shi, M., Wakabayashi, T., Kimura, F.: Handwritten Numeral Recognition Using Gradient and Curvature of Gray Scale Image. In: Proceedings of the Fifth International Conference on ICDAR 1999, Bangalore, pp. 277–280 (1999)
10. Haykin, S.: Neural Networks. McMaster University, Toronto (1994)
11. Lopes, H.F.S., Abe, J.M., Anghinah, R.: Application of Paraconsistent Artificial Neural Networks as a Method of Aid in the Diagnosis of Alzheimer Disease. Journal of Medical Systems 34(6), 1073–1081 (2010)
12. Lopes, H.F.S., Abe, J.M., Kanda, P.A.M., Machado, S., Velasques, B., Ribeiro, P., Basile, L.F.H., Nitrini, R., Anghinah, R.: Improved Application of Paraconsistent Artificial Neural Networks in Diagnosis of Alzheimer's Disease. American Journal of Neuroscience 2(1), 54–64 (2011)
13. Mario, M.C., Abe, J.M., Ortega, N., Del Santo Jr., M.: Paraconsistent Artificial Neural Network as Auxiliary in Cephalometric Diagnosis. Artificial Organs. 34(7), 215–221 (2010)
14. Mori, S., Suen, C.Y., Yamamoto, K.: Historical Review of OCR research and Development. Journals & Magazines 80(7), 1029–1058 (1992)
15. Trier, O.D., Jain, A.K., Taxt, T.: Feature extraction methods for character recognition - a survey. Pattern Recognit. 29(4), 641–662 (1996)
16. Hammerstrom, D.: Adaptative Solutions Inc. Spectrum 30(6), 26–32 (1993)
17. Öksüz, O.: Vision Based Handwritten Character Recognition. M.S. thesis. Bilkent University, Turkey (2003)
18. Javadevan, R., Kolhe, S.R., Patil, P.M., Pal, U.: Automatic processing of handwritten bank cheque images: a survey. IJDAR 15(4), 267–296 (2012)

Substitution Tasks Method for Co-operation

Lidia Dutkiewicz and Ewa Dudek-Dyduch

AGH University of Science and Technology,
al. Mickiewicza 30, 30-059 Krakow, Poland
{lidia,edd}@agh.edu.pl

Abstract. The aim of the paper is to present a novel heuristic optimization method for determining intelligent co-operation at project realization. The method has been named substitution tasks method (ST method). According to the method, a solution is generated by means of sequence of dynamically created local optimization tasks so-called substitution tasks. The method is based on formal algebraic-logical meta model of multistage decision process (ALMM of MDP). The paper presents a formal approach for designing constructive algorithms for co-operation. A general idea of creating substitution tasks is given. To illustrate the presented ideas, a scheduling algorithm for a particular NP-hard problem is given and results of computer experiments are presented.

Keywords: substitution tasks method, optimization of co-operation, project management, algebraic-logical meta model, scheduling problem, multistage decision process.

1 Introduction

The paper is related to the development of a new method within computational intelligence, applied for co-operation algorithms. Intelligent cooperation is defined as an accurate assignment of executors to tasks that form a project. On the one hand, a particular executor is supposed to realize tasks that they are best suited for, on the other, the selection of tasks and their ordering needs to take into account optimal (suboptimal) realization of tasks by other executors.

The aim of the article is to present a new heuristic method for determining intelligent co operation at project realization. The method, named substitution tasks method (ST method), is based on (approximately) decomposing the task of solution generation into a sequence of dynamically created substitution tasks. ST method is based on formal meta model of multistage decision process devised by Dudek-Dyduch [2].

A project is composed of a number of tasks that need to be performed by various executors. The term executor may refer to people and/or machines characterized by a number of differentiating features. Executors should aim towards achieving a common target, which is the realization of a project at minimal costs, and meet restrictions that most often refer to execution time, costs and ordering of particular tasks. The execution time of particular tasks most frequently

A. Bădică et al. (eds.), *Recent Developments in Computational Collective Intelligence*, Studies in Computational Intelligence 513,
DOI: 10.1007/978-3-319-01787-7_10, © Springer International Publishing Switzerland 2014

depends on the assigned executor and the extent of the tasks performed so far (the set of performed tasks). This is because some of the already realized tasks may facilitate or accelerate other task realization. One must determine such an ordering of tasks and assignment of these to particular executors which would minimize the total project cost and meet the restrictions. A lot of combinatorial problems correspond to the co-operation problem presented above, e.g. tasks scheduling on multiple machines. Research in this area are conducted in two main directions: developing algorithms for particular problems and developing general solution methods. The latter includes research devoted to metaheuristics and software tools implementing them. The paper also belongs to the second group of research.

ST method imitates the way in which a human decision-maker deals with a complex problem. When the decision-maker has to make a sequence of decisions in a long time horizon to achieve a specified goal and it is hard to predict long-term results, this decision-maker usually proceeds as described below. Instead of analyzing the whole problem, he or she tries to replace it with a few relatively simple partial problems (intermediate goals). As each partial task has a shorter time horizon, it is easier to determine suitable decisions for it. Furthermore, while solving a given partial task, additional knowledge can be collected and applied in the process. After implementing each single decision, the decision-maker reconsiders resulting situation. Based on the gained information, the partial task is updated (replaced, modified or kept as is). The procedure described above serves as a basis for the heuristic substitution tasks method presented in this paper. ST method can be, therefore, classified as artificial intelligence method.

Proposed method can be used for solving many discrete, dynamic optimization problems. It is useful for control of discrete manufacturing process (e.g. tasks scheduling on multiple machines.), project management and many other combinatorial problems. The paper extends ideas given in [7] [10] [11] [4] [5].

The aim of the paper is 2-fold:

- to present a formal approach for designing constructive algorithms that use special local optimization tasks, so called substitution tasks, and are based on a formal algebraic-logical meta model of multistage decision process (ALMM of MDP),
- to present an application of substitution tasks method for an NP-hard scheduling problem.

2 Algebraic-Logical Meta Model of Multistage Decision Process

An admissible way of project realization, in particular assigning executors to tasks, can be determined with the help of simulation experiments. A single experiment establishes a sequence of decisions related to the assignment of executors (resources) to tasks and task realization ordering. It is impossible to provide a sensible sequence of decisions a priori. It needs to be established in the simulation course. Simulation course of project realization consists in determining a sequence of process states and the related time instances. The new state and its time instant depend on the previous state and the decision that has been realized (taken) then. The decision determines the task to be performed, resources, transport unit, etc. Project realization processes belong to the larger class of discrete process that can be modelled by means of multistage decision process.

Any discrete simulation process that is joined with decision process can be formally presented by meta model of multistage decision process devised by Dudek-Dyduch [2]. Let us recall its definition.

Definition 1. *A multistage decision process (MDP) is a process that is defined by the sextuple* $MDP = (U, S, s_0, f, S_N, S_G)$ *where* U *is a set of decisions,* $S = X \times T$ *is a set named a set of generalized states,* X *is a set of proper states,* $T \subset \mathbb{R}^+ \cup \{0\}$ *is a subset of non-negative real numbers representing the time instants,* $f: U \times S \rightarrow S$ *is a partial function called a transition function, (it does not have to be defined for all elements of the set* $U \times S$*),* $s_0 = (x_0, t_0)$*,* $S_N \subset S$*,* $S_G \subset S$ *are respectively: an initial generalized state, a set of non-admissible generalized states, and a set of goal generalized states, i.e. the states we want the process to reach in the end. The transition function is defined by means of two functions,* $f = (f_x, f_t)$ *where* $f_x: U \times X \times T \rightarrow X$ *determines the next state and* $f_t: U \times X \times T \rightarrow T$ *determines the next time instant. It is assumed that the difference* $\Delta t = f_t(u, x, t) - t$ *has a value that is both finite and positive.*

Thus, as a result of the decision u that is taken or realized at the proper state x and the moment t, the state of the process changes to $x' = f_x(u, x, t)$ that is observed at the moment $t' = f_t(u, x, t) = t + \Delta t$.

Since not all decisions defined formally make sense in certain situations, the transition function f is defined as a partial function. Thus all limitations concerning the control decisions in a given state s can be defined in a convenient way by means of so-called sets of possible decisions $U_p(s)$, and defined as: $U_p(s) = \{u \in U : (u, s) \in Dom f\}$

Because many different types of discrete decision problems can be modeled by means of the above formal model the paradigm $MDP = (U, S, s_0, f, S_N, S_G)$ is in fact a meta model of multistage decision process. Also any possible way of project realization can be formally presented by means of MDP.

At the same time, a fixed MDP (i.e. when U, S, s_0, f, S_N, S_G are defined) represents a set of its trajectories that start from the initial state s_0. Let us recall that trajectory is a sequence of states $\tilde{s} = (s_0, s_1, s_2, \ldots s_i, s_{i+1}, \ldots s_k)$ where $s_{i+1} = f(u_i, s_i)$ and $s_k \in S_G \cup S_N$ or $U_p(s_k) = \emptyset$ It is assumed that no state of a trajectory, apart from the last one, may belong to the set of non-admissible generalized states S_N or may have an empty set of possible decisions $U_p(s)$. Only a trajectory that ends in the set of goal states S_G is admissible. The decision sequence $\tilde{u}$ determining an admissible trajectory is an admissible decision sequence.

In the most general case, sets U and X may be presented as a Cartesian product $U = U^1 \times U^2 \times ... \times U^m, X = X^1 \times X^2 \times ... \times X^n$ i.e. $u = (u^1, u^2, ..., u^m)$, $x = (x^1, x^2, ..., x^n)$. Particular u^i, $i = 1, 2, ..m$ represent separate decisions that must or may be taken at the same time and relate to particular objects in the process (executors, resources, tasks etc.). There are no general limitations imposed on the sets; in particular they do not have to be numerical. The values of particular co-ordinates of a state may be names of individuals (symbols) as well as some objects (e.g. finite set, sequence etc.). The sets S_N, S_G , and U_p are formally defined with the use of logical formulae. Therefore, the complete model constitutes a specialized form of an algebraic-logical model.

The MDP can represent all potential possibilities of a project realization. The knowledge regarding the process (all rules and constraints) is represented by U, S, s_0, f, S_N, S_G. The admissible trajectory corresponds to the admissible project realization.

The task of optimization is to find such an admissible decision sequence $\tilde{u}$ that minimizes a certain criterion Q. The optimization problem is defined by the pair (P, Q), where P represents all the constraints imposed on the process modeled by MDP.

The ALMM of MDP constitutes the basis for defining novel heuristic discrete optimization methods. Based on ALMM of MDP, learning-based method has been proposed and developed [2], [6], [9], [8], [13] as well as a method of production planning in failure modes [14]. Automatic creation of lower bounds for branch and bound method has been also worked out basing on ALMM [1]. ALMM of MDP makes it possible to define mathematic properties of discrete optimization problems. As a result, the proposed heuristic methods and algorithms can be explained and discussed formally. ST method proposed in this article will be also presented in this way.

3 The Idea of the Substitution Tasks Method

Substitution tasks method is a constructive method in which whole trajectories are generated. While generating the solution (i.e. the process trajectory), in each state s of the process a decision is made on the basis of a specially constructed optimization task named substitution task $ST(s)$. The substitution task may be different in each state of the process. Substitution tasks are created to facilitate the decision making at a given state by substituting global optimization task with a simpler local task. After determining the best decision $u^*(s)$, the next process state s' is generated. Then, an automatic analysis of the new process state is performed and, on the basis of information gained, a new or modified substitution task is defined. Thus, in each iteration of the method, computations are performed at two levels:

1. The level of automatic analysis of the process and constructing a substitution task.
2. The level of determining possibly optimal decision for the substitution task and computing the next state.

Substitution task $ST(s) = (P_{ST}, Q_{ST})$, where: P_{ST} - a certain substitution multistage process and Q_{ST} - substitution criterion. In order to highlight that the substitution process is constructed for the state s, we use the notation $P_{ST}(s)$.

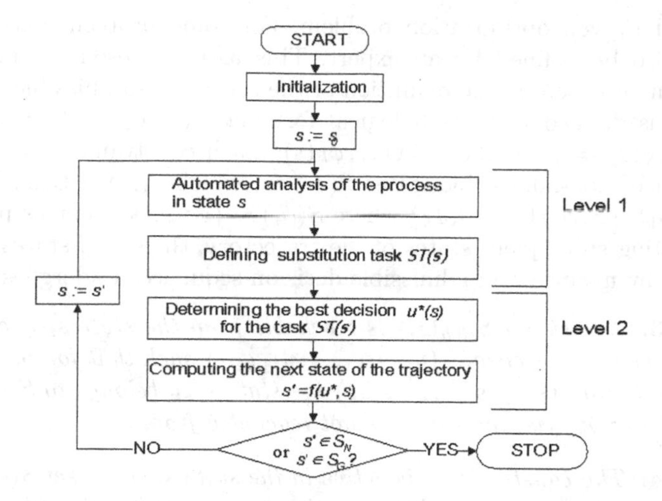

Fig. 1. Schemma of substitution tasks method

Substitution task construction presented in this paper is based upon the concept of so-called intermediate goals.

Definition 2. *Intermediate goal d is defined as reaching by the process, as soon as possible, a certain set of states* S_d.

In scheduling problems, it is most often the case that the subset of states associated with the intermediate goals consists of such states in which the distinguished task or one of the distinguished tasks is completed. The distinguished intermediate goals are used to define the set of final states of the substitution process P_{ST}. Thus for the substitution process a new set of final states $S_{G_{ST}}$ is defined, the initial state s_{0z} is the current state s of the basic process P, whilst the transition function and the sets U, S, S_N are the same as for the basic process P. As a result, the substitution process is defined as follows.

$$P_{ST}(s) = (U, S, s_{0z}, f, S_N, S_{G_{ST}}) \tag{1}$$

It needs to be emphasized that the substitution task $ST(s)$ is for choosing only one single decision in the state s and not for determining a sequence of decisions leading the process P_{ST} from this state to the set of final states $S_{G_{ST}}$.

4 Automatic Analysis and Creation of Substitution Tasks

The following question arises: is it possible to define how substitution tasks should be constructed for any type of problem? The answer is: yes but the problem must be presented by algebraic-logical model (ALM). Based on the analysis, the intermediate goals d_i, $i = 1, 2, \ldots$ are determined in a heuristic way. The rules for goal determining and procedures based on them strongly depend on the given optimization problem. For some problem instances, the goals may also be defined by an expert. This article presents a method for goal determining based on the definition of the set of non-admissible states S_N. The set S_N is defined through a logical formula ϕ_N. $S_N = \{s : \phi_N(s)\}$, in particular: $\phi_N(s) = \phi_1(s) \cup \phi_2(s) \cup \ldots \phi_k(s)$, Each constraint ϕ_i is connected with a subset of non-admissible states $S(\phi_i)$ for which $\phi_i(s)$ is true. Therefore: $S_N(s) = S(\phi_1) \cup S(\phi_2) \cup \ldots S(\phi_k)$ where $S(\phi_i) = \{s : \phi_i(s)\}$ Let us notice that while generating subsequent states of the trajectory, the set of states that may be reachable by means of an admissible decision sequences undergoes changes.

Definition 3. *The subset $S_{Rch}(s_i)$ is reachable from the state s_i if and only if there is a sequence of decisions $(u_i, u_{i+1}, \ldots u_{i+k-1})$ such that for the generated part of the trajectory $(s_{i+1}, s_{i+2}, \ldots s_{i+k})$ the state s_{i+k} belongs to S_{Rch}. If such a sequence does not exist the subset is not reachable from s_i.*

Definition 4. *The constraint ϕ_j is active in the state s_i if the set $S(\phi_j)$ defined through that constraint is reachable from the state s_i. Otherwise, the constraint ϕ_j is inactive in a given state.*

Properties given in definition 4 can easily be verified in a process of automatic analysis as a part of an algorithm based on the method.

There is a large class of problems which are characterized by the following property: if a constraint is no longer active at a certain trajectory state s, then it remains inactive in all further states of any trajectories that starts from the state s. We called this property *a permanent constraint inactivity*. A major part of the known scheduling problems has this property. For the needs of the further discussion, let us assume that the considered optimization problem possesses this property as well. The set of constraints active in a given state defines the subset $S_{NA}(s) \subseteq S_N$ and referred to as active non-admissible set. The more constraints inactive in a given state, the smaller the active set of non-admissible states, and the higher the chances for generating an admissible trajectory. At the same time there is much more freedom in the decision-making process (higher possibility of making decisions which are advantageous in the context of the criterion).

The general idea is to generate a trajectory in such a way as to deactivate certain constraints and at the same time make the subsets of S_N unreachable in an advantageous order. Thus, intermediate goal d_i will be to reach the subset of states S_{d_i}, for which a certain constraint ϕ_i becomes inactive. The set D will be implied by the subset of active constraints that we want to deactivate in a possibly shortest time. The priorities will define the scheduling of eliminating active constraints, that is most favorable order in which the trajectory should reach the subsets of states connected with these goals.

Recall that $ST = (P_{ST}, Q_{ST})$. In order to define the substitution process P_{ST}, it is necessary to define a new set of goal sates $S_{G_{ST}}$. This set is defined on the basis of the distinguished subset of intermediate goals $D_W(s) \subseteq D$. When choosing the subset $D_W(s)$ it is most often the case that priorities of goals are taken into consideration. Also certain additional information can be included, especially related to, for example, the current system state. The number k of intermediate goals selected for the subset $D_W(s)$ can be the same or different at each iteration step. Afterwards, a set of final states $S_{G_{ST}}$ of the substitution process $P_{ST}(s)$ is defined as intersection of all subsets S_{d_i} defined by particular intermediate goals $d_i \in D_W(s)(i = 1, 2, \ldots k)$.

$$S_{G_{ST}} = \bigcap_{d_i \in D_W(s)} S_{d_i} \tag{2}$$

Summing up, construction of a substitution task in a given state s is realized through the following steps: definition of the set of goals D, definition of priorities in the set of goals D, choice of the set $D_W(s)$ of goals for realization, and definition of the set of final states $S_{G_{ST}}$ for the substitution process P_{ST}.

The method of selecting decision u^* is strongly dependent upon the substitution task (see example). Due to the fact that the entire problem is formally presented as an algebraic-logical model, both analysis and substitution task creation methods can be algoritmized.

5 Scheduling Problem with State Depended Resources

To illustrate an application of the model, let us consider a certain collaborative enterprise, where the aim is to develop sophisticated custom software. The development process is connected with the development of some physical product in a client company. The software is already specified and designed, so all its required components are defined. The tasks within the development project (activities) consist of implementing the particular software components. There are deadlines given for some components. They result from arrangements with the client, who wants to use the software in his own product development process.

The software development is performed by a virtual (distributed) team composed of individual programmers. The project manager distributes the components to be implemented among the programmers taking into account their capabilities, efficiency and cost of work. The total cost of development should be minimized and time constraints (deadlines) should be respected.

Activities (implementing the components) have, as usual in the project, precedence constraints. Due to the connection with the physical product development, sometimes there is also a need of preparatory operations (installation of dedicated software, integration with some physical elements). Moreover, some finished components can facilitate (speed up) implementation of other components. Thus it is a problem with state depended resources. It is strongly NP-hard.

5.1 Formal Model of Problem

Let C denotes set of software components and M - set of programmers. The process state at any instant t is defined as a vector $x = (x^0, x^1, x^2, ..., x^{|M|})$, where $|M|$ - number of programmers. A coordinate x^0 describes a set of software components that have been implemented to the moment t. The other coordinates x^m describes state of the m-th programmer, where $m = 1, 2, ..., |M|$. A programmer state specifies which component $c \in C$ he or she is currently implementing, the time remaining to complete the component and additional information about implementation process (preparatory activities, special resources). A state $s = (x, t)$ belongs to the set of non-admissible states S_N if there is a component not implemented yet and its deadline is earlier than t. When we want to emphasize that a component c has deadline, we use a notation $\hat{c}$. Let $\hat{c}_1, \hat{c}_2, \dots \hat{c}_i, \dots \hat{c}_k$ denote k components that have a deadline and $\hat{C}$ - set of these components. The definition of S_N is as follows: $S_N = \{s = (x, t) : (\exists \hat{c}_i \in \hat{C}, \hat{c}_i \notin x^0) \wedge deadline(\hat{c}_i) < t\}$, where $deadline(\hat{c}_i)$ denotes the deadline for the component $\hat{c}_i$. Thus $S_N = S(\phi_1) \cup S(\phi_2) \cup \dots \cup S(\phi_k)$ where $S(\phi_i) = \{s = (x, t) : \hat{c}_i \notin x^0) \wedge deadline(\hat{c}_i) < t\}$ for $i = 1, \dots k$. A state $s = (x, t)$ is a goal state if all the components of the software are implemented. The definition of the set of goal states S_G is as follows: $S_G = \{s = (x, t) : s \notin S_N \wedge (\forall c \in C, c \in x^0\}$.

A decision determines what component each programmer should continue or start to implement at the moment t. Programmer may also stay idle (represented by value 0). Thus, the decision $u = (u^1, u^2, \dots, u^{|M|})$ where the co ordinate u^m refers to the m-th programmer and $u^m \in C \cup \{0\}$. Obviously, not all decisions can be taken in the state (x, t). The decision $u(x, t)$ must belong to the set of possible (reasonable) decisions $U_p(x, t)$.

Based on the current state $s = (x, t)$ and the decision u taken in this state, the subsequent state $(x', t') = f(u, x, t)$ is generated by means of the transition function f. The complete definition of the set of the possible decision $U_p(x, t)$ as well as the transition function f will be omitted here because it is not necessary to explain the idea of the substitution tasks method. The more detailed models of analogical problems are presented in [10] [12].

5.2 Algorithm

Intermediate goals $d_i, i = 1, 2, \dots k$, are a result of S_N subset analysis, analysis of how the decision influences the criterion and the analysis of other subsets of advantageous and disadvantageous states. Depending on the selected intermediate goals, it is possible to propose various algorithms based on substitution task method. Let us consider constraints defining the set S_N. $S_N = S(\phi_1) \cup S(\phi_2) \cup \dots \cup S(\phi_k)$ where $S(\phi_i) = \{s = (x, t) : \hat{c}_i \notin x^0) \wedge deadline(\hat{c}_i) < t\}$ for $i = 1, \dots k$. It is easy to see that the problem satisfies the property of permanent constraint inactivity. By definition of set S_N, intermediate goals $d_i, i = 1, 2 \dots k$ are reaching those subsets S_{d_i}, from which the subsets of non-admissible states $S(\phi_i), i = 1, 2 \dots k$ will not be reachable. This guarantees that the admissible trajectory (solution) will be found. Reaching the goal reduces the subset S_{NA}. Let us notice that the use of formal structure of the algebraic-logical model makes it possible to determine intermediate goals through automatic inference (analysis of logical formulae).

At the same time, the earlier all components with deadlines are implemented the earlier it will not be necessary to choose programmers with high efficiency, but also high costs. As a result, it is possible to take those decisions for which the value of the criterion is lower. Intermediate goal d was then defined as reaching, in the shortest time, the subset of states $S_{d_i} \subset S$, in which the component $\hat{c}_i$ is finished. This subset is formulated in the following way: $S_{d_i} = \{s = (x, t) : \hat{c}_i \in x^0(s)\}$ where $x^0(s)$ is the value of coordinate x^0 in the state s. Let us notice that for this set the constraint $\phi_i(s)$ becomes inactive. The optimization algorithm based on the presented approach is as follows.

Step 1. Setting the initial process state: $s := s_0$

Step 2. Determining intermediate goals
In the given set of software components, components $\hat{c}$ are searched for (that is components with deadlines). Each of them is associated with an intermediate goal d. The number of intermediate goals in the set D is, therefore, equal to the number of active constraints in the state s_0.

Step 3. Calculating priorities of intermediate goals
Each goal d is assigned a priority $p(d)$ which depends on the estimated time slack for accomplishing the component $\hat{c}$ associated with this goal. The slack time is the highest time reserve calculated assuming that the component and all its eventual required predecessors are implemented as soon as possible (by programmers with the highest efficiency). The lower the slack time for the given component the higher the priority of the goal.

Step 4. Determining the set of intermediate goals for realization $D_W(s)$
In this set $|M|$ of intermediate goals with highest priorities is placed (where $|M|$ is the number of programmers). If the set D is smaller than the set M, there is $|D|$ of intermediate goals in the set $D_W(s)$.

Step 5. Determining decision for the given state
In order to make decision $u^*(s)$ in the given state s, it is necessary to define all coordinates of this decision $u(s) = (u^1, u^2, \ldots, u^{|M|})$. The task is partially decomposed by the definition of intermediate goals $d \in D_W(s)$. It is therefore possible to correlate the realization of particular goals with subsequent decision coordinates u^m, which means dedicating these goals to particular programmers. The realization of the goal implies implementing the particular component $\hat{c}$. If some other component(s) are needed to complete prior to component $\hat{c}$, first one is taken as decision for programmer m. If not, obviously, component $\hat{c}$ is assigned. Additionally, if there is a need of some preparatory activities, their cost and time are included.

Step 6. Calculating the next state
Having established the decision $u^*(s)$, we calculate the next process state s' using the transition function $s' = f(u^*, s)$.

Step 7. Verifying stop conditions
If the state s' belongs to goal or non-admissible states, then generating the trajectory will be terminated and the result will be stored. If this is not the case, the next step should be taken.

Step 8. Updating the set D
After determining the next state, it is necessary to update the set D of intermediate goals. What is removed from the set is every goal d, for which the relevant component $\hat{c}$ has been implemented. The number of goals in the set D is always equal to the number of active constraints.

Step 9. Initiating the next algorithm iteration
To begin the next step, it is necessary to set $s = s'$, and then to move to step 3 (updating priorities of intermediate goals).

5.3 Experiments

The algorithm proposed in this paper has been verified through a set of example projects (instances) with a number of activities varying from 20 to 27. The input of the algorithm, for the specified model ALM of the problem, consists of activities and their relations, performers (executors) and resources, cost and duration of activities (dependent on the performer an the resources used). The output is a sequence of decisions assigning activities to performers. The computations were made for the examples with two performers. The presented algorithm generated admissible trajectories for the majority of instances, whilst the calculation time for each of these did not exceed 1 s (for Pentium IV). In order to evaluate the algorithm's efficiency for tested instances, brute-force search has been used. The results are presented in Table 1. Based on this data it can be concluded that the proposed algorithm finds a relatively good solution very quickly.

Table 1. Comparison of the results obtained with the algorithm based on ST Method and the brute-force search algorithm (* - best result obtained until the calculations were stopped)

Identifier of a project	Value of the criterion for ST algorithm	Value of the criterion for brute-search	Error	Calculation time for brute-search
A	14434,50	13880*	3,99%	30h
B	14419,20	13942,80	3,42%	20h 1min 49s
C	15202,30	14777,20	3,99%	29h 1min 55s
D	20030,90	18971,59*	5,58%	30h
E	26844,80	25880,88*	3,72%	30h

6 Conclusions

The paper presents a novel heuristic method for working out co-operation for process realizations, named substitution tasks method (ST method). The method is a construction one, i.e. it generates whole solution (trajectory). The method is based on decomposing the task of solution generation into a sequence of dynamically created local optimization tasks (substitution tasks). The method is based on a general formal model of multistage decision processes (ALMM of MDP), that is given in the paper. Thanks to ALMM of MDP, the way of substitute task creation is proposed and presented formally in the paper.

It is easy to notice that many scheduling problems have a property that we call permanent constraint inactivity and, thus, the metaheuristics proposed in this paper is applicable to them. Especially, the method is very useful for difficult scheduling problems with state depended resources. Managing projects, especially software projects belongs to this class. To illustrate the concept, some NP-hard scheduling problem with state depended resources is considered and the algorithm based on ST method is presented. Results of computer experiments confirm the efficiency of the algorithm.

References

1. Dudek-Dyduch, E.: Control of discrete event processes - branch and bound method. In: Proc. of IFAC/Ifors/Imacs Symposium Large Scale Systems: Theory and Applications, Chinese Association of Automation, vol. 2, pp. 573–578 (1992)
2. Dudek-Dyduch, E.: Formalization and Analysis of Problems of Discrete Manufacturing Processes. Scientific bulletin of AGH UST. Automatics 54 (1990) (in Polish)
3. Dudek-Dyduch, E.: Learning based algorithm in scheduling. Journal of Intelligent Manufacturing 11(2), 135–143 (2000)
4. Dudek-Dyduch, E., Dutkiewicz, L.: Substitution task method for NP-hard scheduling problems. Scientific bulletin of Silesian University of Technology no.1726. Automatics 143, 57–66 (2006) (in Polish)
5. Dudek-Dyduch, E., Dutkiewicz, L.: Substitution Tasks Method for Discrete Optimization. In: Rutkowski, L., Korytkowski, M., Scherer, R., Tadeusiewicz, R., Zadeh, L.A., Zurada, J.M. (eds.) ICAISC 2013, Part II. LNCS, vol. 7895, pp. 419–430. Springer, Heidelberg (2013)
6. Dudek-Dyduch, E., Dyduch, T.: Learning algorithms for scheduling using knowledge based model. In: Rutkowski, L., Tadeusiewicz, R., Zadeh, L.A., Żurada, J.M. (eds.) ICAISC 2006. LNCS (LNAI), vol. 4029, pp. 1091–1100. Springer, Heidelberg (2006)
7. Dudek-Dyduch, E., Fuchs-Seliger, S.: Approximate algorithms for some tasks in management and economy. System, Modelling, Control 1(7) (1993)
8. Dudek-Dyduch, E., Kucharska, E.: Learning method for co-operation. In: Jędrzejowicz, P., Nguyen, N.T., Hoang, K. (eds.) ICCCI 2011, Part II. LNCS, vol. 6923, pp. 290–300. Springer, Heidelberg (2011)
9. Dudek-Dyduch, E., Kucharska, E.: Optimization learning method for discrete process control. In: ICINCO 2011: Proceedings of the 8th International Conference on Informatics in Control, Automation and Robotics, vol. 1 (2011)
10. Dutkiewicz, L.: Two-Level Algorithms for Optimization of Production Processes with Resources Depending on System State. PhD thesis (2005) (in Polish)
11. Dutkiewicz, L.: Two-level algorithms for scheduling problem. In: Rutkowska, D., et al. (eds.) Selected Problems of Computer Science, pp. 139–144. Academic Publishing House EXIT, Warsaw (2005)
12. Dutkiewicz, L., Kucharska, E.: Model of scheduling problem with state dependent resources. Semiannual: Automatics, AGH University of Science and Technology 9(1-2), 67–77 (2005)
13. Kucharska, E., Dudek-Dyduch, E.: Extended Learning Method for Designation of Co-operation. Transactions on Computational Collective Intelligence (to appear)
14. Sękowski, H., Dudek-Dyduch, E.: Knowledge based model for scheduling in failure modes. In: Rutkowski, L., Korytkowski, M., Scherer, R., Tadeusiewicz, R., Zadeh, L.A., Zurada, J.M. (eds.) ICAISC 2012, Part II. LNCS, vol. 7268, pp. 591–599. Springer, Heidelberg (2012)

Modular Localization System for Intelligent Transport

Michal Mlynka, Peter Brida, and Juraj Machaj

University of Zilina, Faculty of Electrical Engineering,
Department of Telecommunications and Multimedia, Univerzitna 1, 010 26 Zilina, Slovakia
{michal.mlynka,peter.brida,juraj.machaj}@fel.uniza.sk

Abstract. Positioning is a very important feature due to development of different Location Based Services (LBS). Ubiquitous positioning is the goal that is quite hard to achieve using one localization system. Global Navigation Satellite Systems (GNSS) are widely used in large number of applications. However, they are usable only if there is a line of sight to satellites. This paper deals with Modular Localization System that utilizes existing radio networks infrastructures together with GNSS. The modularity means use of multiple independent technologies that allow determining geographical position in different geographical environments. Modular Localization System which has been designed uses modules based on Global System for Mobile Communications (GSM), Global Positioning System (GPS) and IEEE 802.11b/g standard (Wi-Fi) to estimate the position of mobile device. Smartphones with the Android operating system were chosen as target devices which position will be estimated.

Keywords: Localization, Localization system, Modular localization system, Positioning, Ubiquitous positioning, Intelligent transport system, Android, fingerprinting.

1 Introduction

Localization systems have been used in the many sectors of our life e.g. military, industrial, agricultural and commercial sectors. Application solutions that use these systems are various navigation systems, tracking systems or searching systems (e.g. in warehouses). Wide application possibility of these systems requires their deployment in diverse environments.

There are many localization systems and each has its own advantages and disadvantages. Global Navigation Satellite Systems (GNSS), e.g. Global Positioning System (GPS) offers great coverage area with good accuracy in the outdoor environment. Unfortunately, these systems are not applicable in indoor environment and their accuracy in the urban environment is not as high as in open outdoor environment due to multipath propagation and obstacles in line of sight [1]. On the other hand, localization systems based on the radio fingerprinting method appear to be the most suitable for indoor and urban environment [2]-[5].

There are some systems that allow localization in multiple environments [6]-[8]. These systems require extra infrastructure which is often impractical and financially

A. Bădică et al. (eds.), *Recent Developments in Computational Collective Intelligence*, Studies in Computational Intelligence 513,
DOI: 10.1007/978-3-319-01787-7_11,

demanding. For these reasons a modular localization system is propose in this paper. This system is designed for use in a multiple environments by using the existing infrastructure and widespread smartphones. This function is achieved by its individual modules.

The paper is structured as follows. Section 2 deals with the architecture of proposed system. Section 3 contains information about used localization methods. Software solution for a mobile device is described in the Section 4. Test scenario is depicted in the Section 5. The Section 6 concludes the paper and presents some ideas for the future work.

2 Modular Localization System

On the basis of the previous research, the modular localization system was developed [8]. Proposed modular localization system is designed to maximize the coverage of the localization service. This was reached by utilization of the different radio networks which work with the different radio frequency. The example is shown in the Fig. 1.

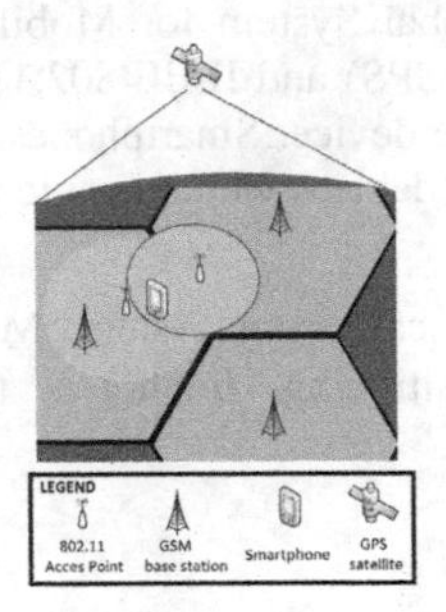

Fig. 1. Example of modular localization system

Logical model of the system is depicted in the Fig. 2. Model of the proposed system consists of the three layers. The lowermost layer consists of the individual localization modules. Openness of the system has been ensured by this layer. Openness means that the system has the ability to add new modular localization blocks e.g. existing GNSS or modules based on radio networks.

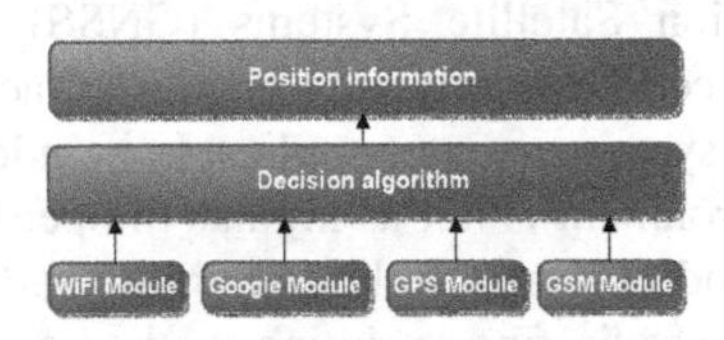

Fig. 2. Logical model of Modular localization system

Smartphone with Android platform has been chosen as a device which position will be estimated. Smartphone with Android is the world's most popular mobile platform. The application of the positioning system has been developed through the Android software development kit (SDK) [10] – [12]. Almost all Android smartphones have integrated GSM, GPS and Wi-Fi chipsets. Based on these facts, four modular localization modules have been implemented. These modules are described in Section 3.

The second layer includes the decision algorithm. The task of this layer is to determine which position estimate will be provided to the user. This decision is based on predetermined criteria. Flowchart of the used decision algorithm is shown in the Fig. 3.

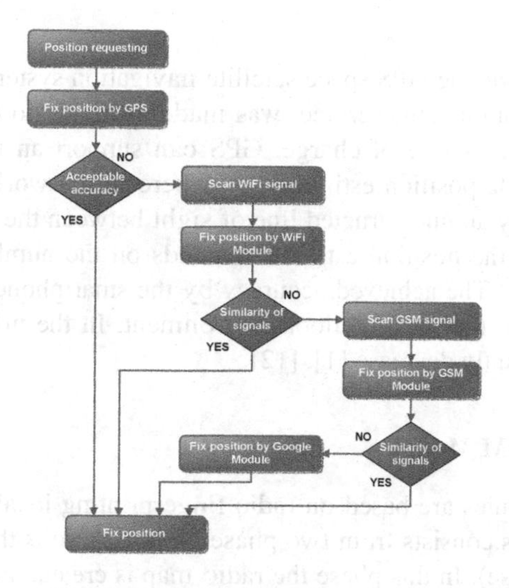

Fig. 3. Diagram of decision algorithm

The decision algorithm has been designed to work with multiple position information. Role of the algorithm is to determine which position information is the most reliable. If the position determined by GPS has acceptable accuracy, it is rated as the most reliable. The acceptable accuracy of GPS module can be set by user. In general, today's smartphones offer GPS accuracy of 4 m in open area. In the dense urban area the positioning error can rise up to 20 m or even more. When accuracy is not acceptable or there is no GPS signal, the algorithm prefers the position information obtained by Wi-Fi localization module. If it is not possible to estimate position of the mobile device using the Wi-Fi module, position information obtained by the GSM localization module is used. There may be situations when both Wi-Fi and GSM localization modules cannot determine the position estimate (e.g. there is no similarity between the actual measured signals and signals stored in the database). In this case, the Google module can be used.

The top layer is Position information. This layer ensures view of the estimated position of user on the map. Also, it can give all radio signals and position information which has been obtained by Decision algorithm layer, to the user. These can be used for environment analysis during the deployment of the localization system.

3 Localization Modules

The system consists of the several localization modules as shown in the Fig. 2. These modules are designed to ensure ubiquity. The particular localization modules are described in the following paragraphs.

3.1 GPS Module

GPS module is based on the GPS space satellite navigation system that provides location and time information. This service, was made available to civilians in 1996 for navigation purposes, it is free of charge. GPS can support an unlimited number of users, and may provide position estimates anywhere in the world. To obtain a location, there is necessary an unobstructed line of sight between the receiver to the satellite. The accuracy of the position estimate depends on the number of used satellites and satellite geometry. The achieved accuracy by the smartphone GPS chipset can be in the range of 4 m in the open outdoor environment. In the urban environment the accuracy can significantly decrease [1], [12].

3.2 Wi-Fi and GSM Modules

Wi-Fi and GSM modules are based on radio fingerprinting localization method. Fingerprinting algorithms consists from two phases. First phase is the offline phase (also called calibration phase). In this phase the radio map is created and stored in the database on the localization server. The second phase is called online phase, in this phase position of mobile device is estimated.

Offline Phase
Area where localization services will be offered is divided into small cells. Each cell is represented by one reference point. Reference points are represented by geographic coordinates. Information about Received Signal Strength (RSS) values from all APs (Access Points) in range are measured at each reference point. Element of radio map has the form:

$$P_j = (N_j, \alpha_{ji}, \beta_{ji}, \theta_j), j = 1,2..., m; i = 1,2..., n, \quad (1)$$

where N_j is number of j-th reference point, m is the number of all reference points, i is number of AP, n is the number of all APs, α_{ji} is the vector of RSS values, β_{ji} stands for the identifier of APs and parameter θ_j obtains additional information which can be used during the localization phase.

Values β_{ji} are tagged by Media Access Control (MAC) address and Cell identity (CID) for Wi-Fi and GSM networks, respectively [2], [3], [13]-[15].

Online Phase

During the online phase the server uses a deterministic nearest neighbor algorithm to estimate the location of mobile devices. Actual measured RSS values received by smartphone are compared with the values P_j stored in the database using the Euclidean distance. Euclidean distance represents the shortest distance between two vectors in Cartesian coordinate system and is defined by:

$$d_{Eij} = \sqrt{(\sum_{k=1}^{n} a_{ik} - b_{jk})^2} \tag{2}$$

where n is number of elements in vector, a_{ik} represents k-th element of vector A and b_{jk} represents k-th element of vector B. Position of the reference point with the smallest Euclidean distance is considered as the estimated position [2], [3], [13]-[15].

3.3 Google Module

Android SDK includes localization library which offers localization of a mobile device by network provider function. This function determines the location of mobile device based on availability of cell tower and Wi-Fi AP. Results are retrieved by mean values of a network lookup. This module does not provide high accuracy. On the other hand, this module provides localization in unknown urban environment [8], [12].

4 Developed Software Solutions

For our experiment an application to Android smartphone has been developed. With this application, the user can create radio maps of Wi-Fi and GSM signals. Later, based on this map, the position of a user may be estimated using Wi-Fi or GSM localization modules. Real position on the map, which is needed for offline phase, can be specified by touch of user finger or given in the World Geodetic System 1984 (WGS 84) coordinates as well. The application enables to calculate distance between points. Currently measured data can be stored to the text file for later analysis [10], [12], [16], [17].

The application can offer multiple position estimates. This feature allows investigating the accuracy of the individual modules in different environment. The red color shows the position estimate obtained by GPS, blue by Google network provider, green by Wi-Fi, cyan by GSM. Information about obtained positions is shown in top left corner as is showed in the Fig. 4. It is also possible to monitor information about the current GSM and Wi-Fi radio signal.

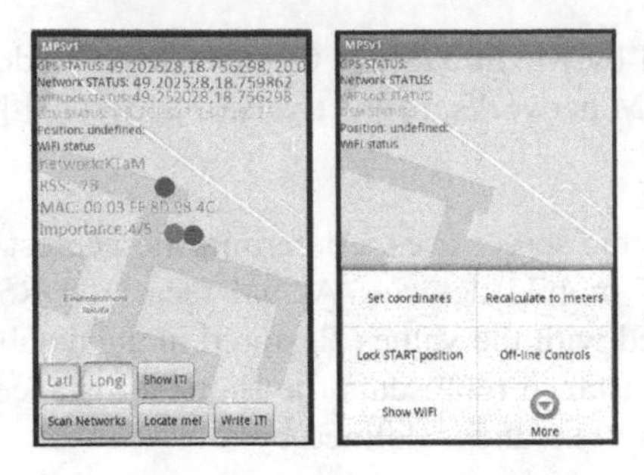

Fig. 4. The screen of developed application

As outdoor maps, OpenStreetMap (OSM) by MapQuest Android API has been used. These maps are open and freely available [16].

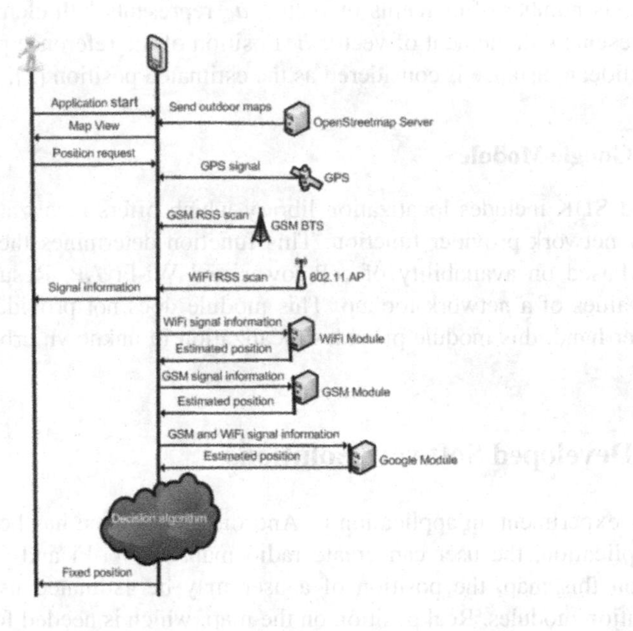

Fig. 5. Sequence diagram – fixed position by smartphone

In the Fig. 5 sequence diagram is shown. This diagram describes how the position of mobile device is estimated by the proposed modular localization system in the time domain.

5 Experimental Scenario

Experimental scenario was performed at the University of Zilina campus. As shown in the Fig. 6, area near the buildings was chosen. In this area poor GPS coverage was expected.

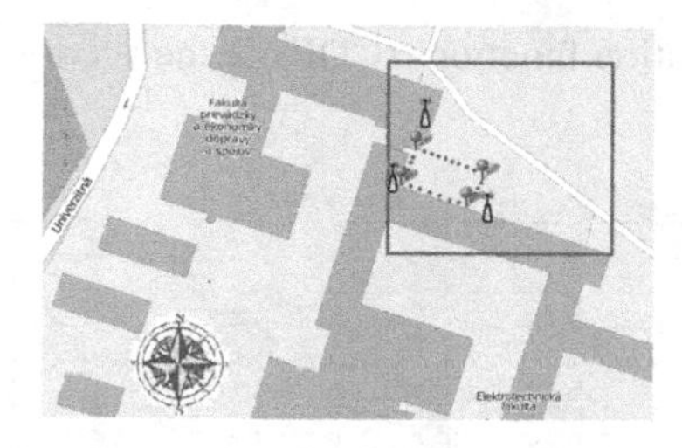

Fig. 6. Experiment area – University of Zilina

Examination area has size of 22x16 meters. Measurements during the offline phase were performed in a grid, with points spaced 2 m apart. Existing radio infrastructure with three added AP was used. 18 APs and 11 BTSs were totally used. Measurements were performed using smartphone HTC Legend. The Fig. 7 depicted how radio map was created. The first, geo-points in chosen area were selected. These geo-points were targeted by Trimble VX [18]. The chosen method does guarantee targeting points with accuracy of 4 cm. In each of the targeted points with poor GPS signal, Wi-Fi and GSM radio signals from nearest APs and BTSs were measured. These measurements were sent to the localization server and stored in the radio map database.

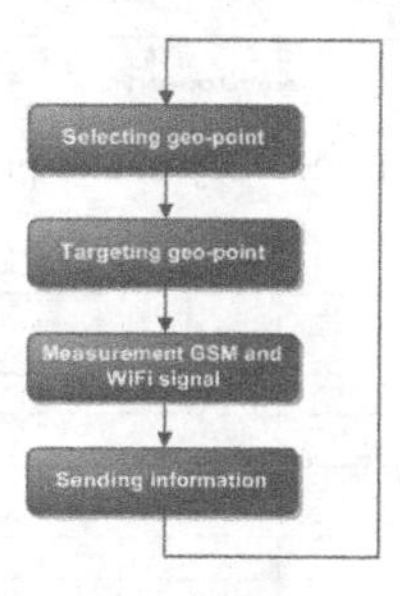

Fig. 7. Process of radio maps creating

After creating the radio maps, accuracy of the localization modules were tested. To evaluate the accuracy of individual localization modules 100 position estimates were performed. Accuracy of a given localization module was obtained as the distance between the real (positions obtained by Trimble VX) and the estimated positions. This distance has been obtained by Vincenty formula. Vincenty formula is commonly used in the geodesy to calculate the distance between two geo-points in WGS 84 system. Mean values of localization error for the individual modules are shown in Tab. 1.

Table 1. Average localization error of the different modules

Module	GPS	Wi-Fi	GSM	Google
Accuracy [m]	26.31	4.82	5.32	69.89

The cumulative distribution functions (CDF) of the accuracy of individual modules are shown in Fig. 8-11.

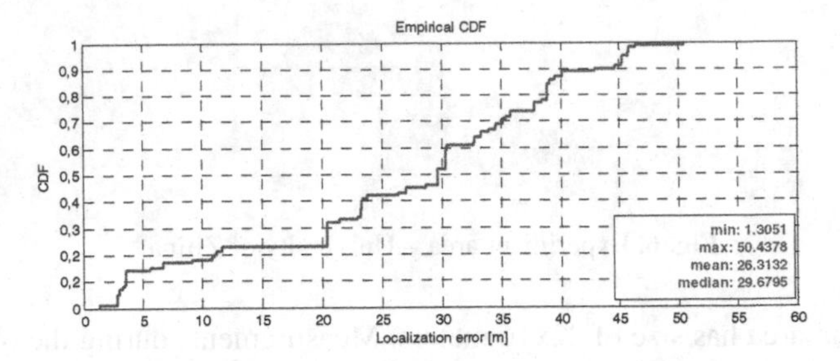

Fig. 8. CDF accuracy of GPS module

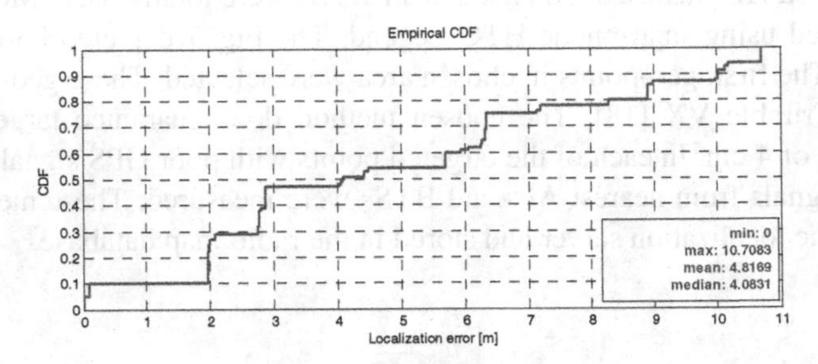

Fig. 9. CDF accuracy of Wi-Fi module

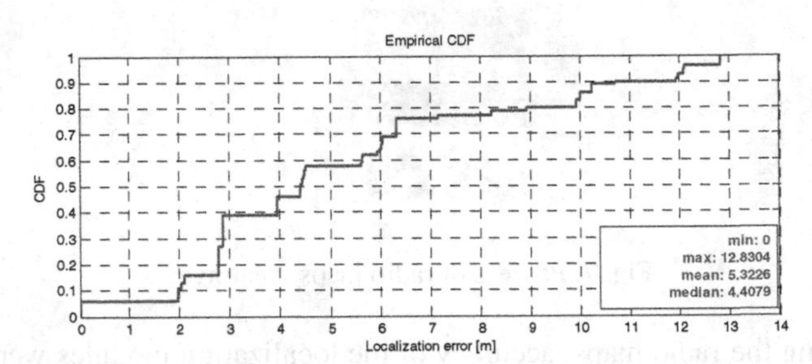

Fig. 10. CDF accuracy of GSM module

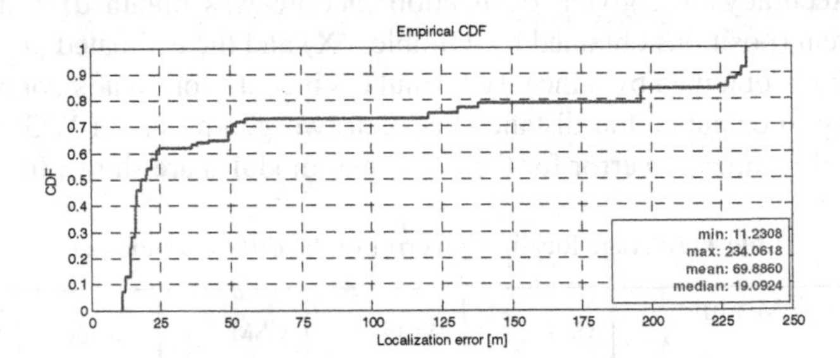

Fig. 11. CDF accuracy of Google module

As shown in Fig. 8-11, GPS accuracy in chosen area wasn't acceptable. GSM and Wi-Fi modules offer better accuracy and they seem to be more suitable. This phenomenon was due to good radio infrastructure. Google module doesn't offer good accuracy but it is usable without needs of offline phase.

6 Conclusion and Future Works

From results shown in this paper it is clear that localization by GNSS (e.g. GPS) is not always the best solution. On the basis of achieved results it can be assumed that Wi-Fi and GSM modules offer better accuracy near the buildings and are important part of modular localization system. Of course, this accuracy depends on the network infrastructure in examined environment, used algorithms, propagation conditions, etc. Google module does not offer good accuracy but this module allows localization in situation when there are no GPS signals and radio maps for other modules are not created. Proposed modular localization system offers higher accuracy and the ability to estimate position in a diverse environment. Whereas that system uses the existing infrastructure, introduction of this system is not financial expensive.

Proposed system increase the accuracy of the position estimates, but the most important is the fact that the system allows localization in areas where GNSS fails. Openness and modularity enable localization in the both outdoor and indoor environment simultaneously.

For future work there is idea to implement new localization modules. New localization module can be represented by existing satellite navigation systems (e.g. Galileo, GLONASS) or by localization based on the utilization of radio networks e.g. DVB-T, FM radio, etc. The introduction of new modules may not require hardware modifications.

Another part which offers improvement the accuracy of the estimated position is decision algorithm. There can be added logic which will work with prediction information (e.g. Kalman Filter).

Acknowledgements. This work has been partially supported by the Slovak VEGA grant agency, Project No. 1/0394/13 and by

„Broker centre of air transport for transfer of technology and knowledge into transport and transport infrastructure ITMS 26220220156"

We support research activities in Slovakia/Project is co-financed by EU

The authors would also like to thank Ing. Juraj Mužík, PhD. and Ing. Andrej Villim, University of Žilina, Faculty of Civil Engineering, Department of Geodesy, Žilina, Slovakia, for their help to target geo-points with Trimble VX.

References

1. U.S. GOVERMMENT: Global Positioning System Standard Positioning Service Signal Specification (2012), http://www.gps.gov/technical/ps/1995-SPS-signal-specification.pdf
2. Bahl, P., Padmanabhan, V.N.: RADAR: An In–Building RF–based User Location and Tracking System. In: Proceedings of the IEEE Infocom, vol. 2, pp. 775–784. Mallows (2000)
3. Krishnan, P., Krishnakumar, A.S., Wen-Hua, J., Mallows, C.: A system for LEASE: location estimation assisted by stationary emitters for indoor RF wireless networks. In: INFOCOM 2004: 23rd Annual Joint Conference of the IEEE Computer and Communications Societies Gamt, vol. 2, pp. 1181–1190 (2004)
4. Cipov, V., Dobos, L., Papaj, J.: Cooperative Trilateration-based Positioning Algorithm for WLAN Nodes Using Round Trip Time Estimation. Journal of Electrical and Electronics Engineering 4(1), 29–34 (2011)
5. Klingbeil, L., Romanovas, M., Schneider, P., Traechtler, M., Manoli, Y.: A Modular and Mobile System for Indoor Localization. In: International Conference, Indoor Positioning and Indoor Navigation (IPIN), pp. 1–10 (2010)
6. Rabinowitz, M., Spilker Jr., J.J.: A New Positioning System Using Television Synchronization Signals. IEEE Transactions Broadcasting 51, 51–61 (2005)
7. EKAHAU, INC - Wi-Fi Tracking Systems, RTLS and WLAN Site Survey (2012), http://www.ekahau.com/
8. Choi, J.S., Hyun, L., Elmasri, R., Engels, D.W.: Localization Systems Using Passive UHF RFID, INC, IMS and IDC. In: 5th International Joint Conference NCM 2009 pp. 1727–1732 (2009)
9. Benikovsky, J., Brida, P., Machaj, J.: Proposal of user adaptive modular localization system for ubiquitous positioning. In: Pan, J.-S., Chen, S.-M., Nguyen, N.T. (eds.) ACIIDS 2012, Part II. LNCS, vol. 7197, pp. 391–400. Springer, Heidelberg (2012)
10. Google Inc. Android Developers (2012), http://developer.android.com/
11. Behan, M., Krejcar, O.: Open Personal Identity as a Service. In: Zgrzywa, A., Choroś, K., Siemiński, A. (eds.) Multimedia and Internet Systems: Theory and Practice. AISC, vol. 183, pp. 199–207. Springer, Heidelberg (2013)
12. Darcey, L., Conder, S.: Android Wireless Application Development, 3rd edn. Addison-Wesley Professional (2012)
13. Machaj, J., Brida, P.: Performance Comparison of Similarity Measurements for Database Correlation Localization Method. In: Nguyen, N.T., Kim, C.-G., Janiak, A. (eds.) ACIIDS 2011, Part II. LNCS, vol. 6592, pp. 452–461. Springer, Heidelberg (2011)
14. Brida, P., Machaj, J., Benikovsky, J.: WLAN Based Indoor Localization System. ElektroRevue 2, 15–21 (2011)
15. Reddy, S., Bagaria, A., Aggarwal, D., Arora, N.: Analysis of Fingerprint Data in Cellular Networks: An Android Application Case Study. In: IPCSIT 2012, vol. 47, pp. 213–218 (2012)
16. MapQuest, Android Maps API (2012), http://developer.mapquest.com/web/products/featured/android-maps-api
17. Vincenty formula for distance between two Latitude/Longitude points (2013), http://www.movable-type.co.uk/scripts/latlong-vincenty.html
18. Trimble Navigation Inc. Trimble VX (2013), http://www.trimble.com/3D-laser-scanning/vx.aspx?dtID=technical

The Estimation of Accuracy for the Neural Network Approximation in the Case of Sintered Metal Properties

Jacek Pietraszek[1], Aneta Gądek-Moszczak[1], and Norbert Radek[2]

[1] Institute of Applied Informatics, Cracow University of Technology,
Al. Jana Pawla II 37, 31-864 Kraków, Poland
pmpietra@mech.pk.edu.pl, aneta.moszczak@gmail.com
[2] Centre for Laser Technologies of Metals, Technical University of Kielce,
Al. Tysiąclecia Państwa Polskiego 7, 25-314 Kielce, Poland
norrad@tu.kielce.pl

Abstract. The spatial structure of sintered metal powders is described by many qualitative and quantitative micro-geometrical properties. The statistical approach based on univariate and multivariate distributions is very useful for consistent and objective description of such structures. It provides information appropriate for a whole population of sinters, not only particular specimen. Empirical distributions of quantitative properties obtained from the image analysis are very irregular and for this reason inconvenient for further numerical simulations. The smoothing of these distributions is required for data conditioning and preprocessing however, the use of simple regression techniques is limited due to the strict lower and upper bound of cumulative distribution function. Authors propose to use a multilayer perceptron as a non-parametric regression model because of its the well-known smoothing properties. The technical application of such model requires additionally providing of the confidence band or any equivalent measure of uncertainty. The highly non-linear structure of the neural network model makes not possible to use typical linear techniques to estimate variance. Additionally, the simple confidence band estimation leads to non-physical values of the cumulative distribution function: lower than 0 or greater than 1. Authors propose to avoid such difficulties by two methods. Firstly, the lower and upper bound limitation are satisfied by the logit transformation which projects the range [0, 1] into unlimited real range. Secondly, the variance of the neural network model is estimated by jackknife estimator. The article presents such approach with preliminary attempt to an automated data processing by ADCIS Aphelion image analysis software and STATSOFT Statistica data analysis software. The almost full automation of the process is required by materials science engineers due to the lack of the sufficient data processing knowledge and skills. Both software systems provide suitable embedded programming environments: C# for Aphelion and Visual Basic for Statistica. The proposed approach has been tested on the example of pore size distribution in sintered stainless steel AISI 434L.

Keywords: neural network approximation, accuracy estimation, jackknife, cumulative distribution function, logit transformation, sintered metal powders.

A. Bădică et al. (eds.), *Recent Developments in Computational Collective Intelligence*,
Studies in Computational Intelligence 513,
DOI: 10.1007/978-3-319-01787-7_12,

1 Introduction

The analyzed raw data describe the empirical distribution function with small number of classes. The data set contains pairs of two variables:

– independent, numerical with values that are middle points of classes,
– dependent, numerical with random values that are cumulative relative frequencies of observations.

The empirical cumulative distribution function is a monotonic step function while the histogram shows a large irregularity. For numerical simulation purposes, it is useful to create a smooth model of the empirical distribution function with confidence bands. Direct use of the neural network model provides the required smoothness. Unfortunately, two problems appeared. Firstly, the cumulative distribution function has values in the limited range: the lower bound is equal 0 and the upper bound is equal 1. The mean value predicted by the model and the confidence bands must satisfy such limitations. Secondly, the multilayer perceptron is a highly non-linear function and it makes not possible to apply typical methods of the linear algebra to estimate confidence bands.

The following sub-chapters describe in details the approach proposed as a solution: the neural network model as an approximator, the logit space as a support for the output of the model and jackknife sub-sampling for estimation of the output variance.

1.1 The Logit Space Eliminates Meaningless Predictions

The measure of probability varies in the range from 0 to 1. Typical probability functions: density function and cumulative distribution function shows areas of asymptotic saturation near 0 (density function) or 0 and 1 (cumulative distribution function). Such limited range of variability and asymptotic characteristics cause the fundamental problem with a prediction of meaningless values. Typical approximation models have virtually unlimited range of variability. This behavior can lead to exceeding the allowable ranges and the loss of meaningfulness.

Such risk may be avoided by two-step transformation. In the first step, the probability p is transformed into *odds* with formula:

$$\text{odds}(p) = \frac{p}{1-p} \ , \tag{1}$$

and re-transformed with inverse formula:

$$\text{p}(odds) = \frac{odds}{1+odds} \ . \tag{2}$$

The odds is ranged from 0 to positive infinity: odds of 0 maps into probability of 0 and odds of positive infinity maps into probability of 1. In the second step, the odds is transformed into *logit* by natural logarithm:

$$\text{logit}(odds) = \ln(odds)\ . \quad (3)$$

The inverse transformation is trivial:

$$\text{odds}(logit) = \exp(logit)\ . \quad (4)$$

The logit is ranged from negative infinity to positive infinity: logit of negative infinity maps into odds of 0 and logit of positive infinity maps into odds of positive infinity.

The full protection from prediction of meaningless values may be achieved through a combination of both transformation into one. Such transformation maps probabilities from the interval [0,1] into the logit space which variability is equal to the extended $\mathbf{R}^1$ i.e. $\mathbf{R}\cup\{-\infty,\infty\}$ The logit was introduced by Berkson [1] with the transformation:

$$\text{logit}(p) = \ln\left(\frac{p}{1-p}\right) \quad (5)$$

and re-transformation formulas:

$$\text{p}(logit) = \frac{\exp(logit)}{1+\exp(logit)}\ . \quad (6)$$

and heavily utilized in the logistic regression models [2].

This transformation provides sufficient protection against exceeding the limits, because:

- the limit of probability as logit approaches negative infinity is 0,
- the limit of probability as logit approaches positive infinity is 1.

Such mapping is bijective (one-to-one correspondence) and it allows to locate the whole computational procedure, mean and variance prediction, in the logit space and to return to probability space with final results.

1.2 Jackknife Allows to Estimate Variance of Non-linear Model

Non-parametric models do not have a *priori* imposed functional form of the regression relationship, and their design is adaptively controlled by available data. Such approach allows a much better fit of prediction to the raw data. In most cases, the determination of non-parametric models (neural network, non-parametric empirical likelihood, finite element methods etc.) is computationally expensive, so it is advisable to seek the most cost-effective use of the procedures for the identification of such models. In the absence of known in advance functional formula, there is not possible to determine analytically probability distributions of variables and their confidence bands.

Numerical determination of such distributions is possible by using Monte Carlo methods. Bootstrap method is usually used. It allows one to get the entire probability distribution of the output variable. However, this involves the need of multiple (from a few hundred to several thousand times) identification of auxiliary non-parametric models, which can be very costly in time and computationally. If requests are limited to a variance of the output variable, it will allow implementation of the subsampling procedure *jackknife* [3]. It requires the use of a relatively small and acceptable (from a few to a few dozen times) number of identification of auxiliary non-parametric models. At the same time knowledge of variance allows for a reliable estimation of the uncertainty, although the exact form of the distribution is not known.

Taking into account that the neural network model is so non-linear and analytical estimation of confidence bands is practically impossible, the jackknife method appears to be correct approach to the variance estimation problem.

The explicit general nonlinear approximation model has the following formula (eq.7):

$$y_i = f(x_i, \beta) + \varepsilon_i, \quad i = 1, \ldots, n \tag{7}$$

where β is a q-vector of unknown parameters (e.g. weights and biases in neural networks), f is a known nonlinear mapping, x_i are p-vectors, ε_i are random errors with $E(\varepsilon_i|x_i) = 0$. Typical estimators of weights and biases β in neural network regression problems are obtained in supervised approach. The estimators $\hat{\beta}_{LS}$ are taken as least squares estimator (eq.8):

$$L_n(\hat{\beta}_{LS}) = \min L_n(\gamma), \quad \gamma \in \mathrm{B} \tag{8}$$

where B is a set of all possible values of β and criterion L_n is of formula (eq.9):

$$L_n(\gamma) = \frac{1}{2}\sum_{i=1}^{n}[y_i - f(x_i, \gamma)]^2 \tag{9}$$

If the parameter of interest ϑ is defined as a given function g of parameters β, then the LSE estimator of ϑ is (eq.10):

$$\hat{\vartheta}_{LS} = g(\hat{\beta}_{LS}) \tag{10}$$

and jackknife variance estimator of $\hat{\vartheta}_{LS}$ is (eq.11):

$$\upsilon_{JACK} = \frac{n-1}{n}\sum_{i=1}^{n}\left(\hat{\vartheta}_{LS,i} - \frac{1}{n}\sum_{j=1}^{n}\hat{\vartheta}_{LS,j}\right)^2 \tag{11}$$

If g (eq.10) is defined as f (eq.7) at arbitrary x i.e. (eq.12):

$$g(\beta) = f(x, \beta) \tag{12}$$

then ϑ is mean value of f at x. With this assumption, $\hat{\vartheta}_{LS}$ and υ_{JACK} are jackknife estimators of the mean and the variance of f at any arbitrary x (eq.13):

$$\begin{array}{ccc} E\left(\hat{\vartheta}_{LS}\right) & \xrightarrow[n\to\infty]{} & E\left(y|x\right) \\ E\left(\upsilon_{JACK}\right) & \xrightarrow[n\to\infty]{} & \mathrm{var}\left(y|x\right) \end{array} \tag{13}$$

2 Case Study in Material Science

The most significant performance characteristics for friction pairs are the microscopic geometric properties of materials and therefore of utmost importance. A construction of a simple regression model (usually linear) for mean value is a typical approach to the problem of characteristic prediction. Unfortunately, due to the stochastic nature of the phenomena and non-normal distributions involved [4], the estimation of prediction accuracy is rather difficult and very often omitted. The authors investigated the bootstrap approach to such problem [5] as well as the impact of sampling window size [6].

In the mentioned case, the effect of additives on the quality of sintered material is studied. The microstructure properties of the pores in sintered products are the criteria of quality assessment [7, 8]. As a research materials were used the images of the tested materials, obtained by camera integrated with metallography light microscope and a computer. Grabbed images were processed [9, 10] by a specialized software for image processing and analysis ADCIS Aphelion v.3.2. Various quantitative properties are identified for pores, among others their area and circularity.

These properties show non-homogenous distributions described by cumulative empirical distribution and associated histograms. To predict quality of sinters produced with different additives, there is necessary to build a model which allows to simulate distributions of geometric features together with estimates of uncertainty. Unavoidable uncertainties associated with settings in the process of pressing and sintering lead to the dispersion of characteristics for obtained sinters. Reducing the dispersion of estimates would be possible by repetitions (replications) i.e. realization of compacting and sintering with same composition and at the same nominal settings.

However, multiple runs of compacting and sintering introduces to the experiment additional block factors associated with individual differences in mixing additives and realization of compacting and sintering, even at the same nominal settings of the control factors. Assessment of impact for block factors would require significant expansion of the number of completed samples and metallographic analyzes, thus the implementation of perturbation methods based on subsampling allows to achieve a significant reduction in time and cost of the work.

2.1 Raw Data

The sample was made from sintered ferritic stainless steel powder AISI 434L obtained from Höganäs Corporation [11] doped with Mn and Ni. The typical shape form of a grain is presented on Fig.1.

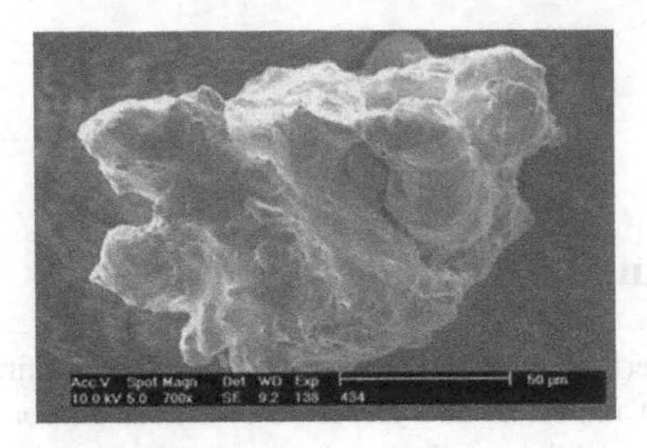

Fig. 1. Typical shape form for a grain of AISI 434L stainless steel powder

The sample was cut and a metallographic specimen was made. The specimen was scanned (Fig.2) by a camera (Olympus model DP-25) mounted on a metallographic microscope (Nikon model Eclipse E400) and acquired into the image analysis software (ADCIS Aphelion).

The data set obtained from the software contains pairs: size threshold in pixels and number of pores equal or smaller than the threshold. The data set was trimmed at area size of 32 pixels due to a large number of artifacts resulting in over-representation of very small pores. The source data are presented in Tab.1.

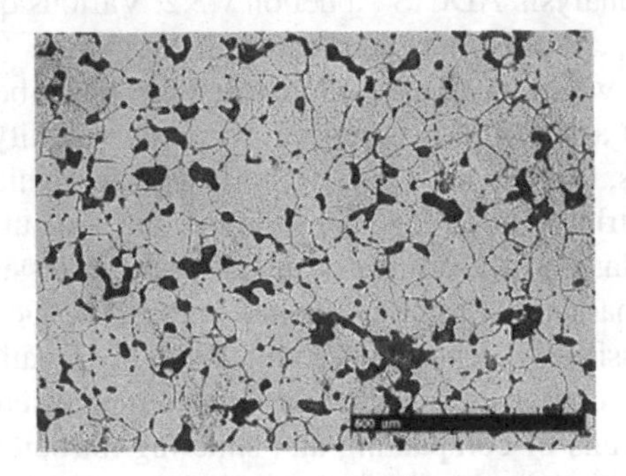

Fig. 2. The metallographic specimen of sintered sample

Table 1. The trimmed data set of pores area population

Area [pixels]	32	33	43	62	93	123	188	265
Count	0	105	1000	2000	3000	3500	4000	4250
Area [pixels]	634	1347	1428	1439	2009	2288	2854	2931
Count	4500	4538	4539	4540	4541	4542	4543	4544

Two pairs in Table 1, (32; 0) and (2931; 4544), describe lower and upper bound of the population. The logit transformation maps these pairs into non-numeric values of negative and positive infinity. Due to this fact the first pair was modified to the count value of 1 and surrogate total count 4545 was introduced replacing 4544. Such approach allows to avoid these infinities.

2.2 Neural Network Approximation

Next, the trimmed data set was processed by neural networks approximators in Statsoft Statistica Automatic Neural Networks module [12].

Due to the size of available data set, three neural networks were considered: 1-2-1, 1-3-1 and 1-4-1 with tanh-tanh activation functions. The pair of activation functions was selected due to the best fitting.

The data were divided into training, validation and testing sets randomly in proportions 70%:15%:15% respectively. The proportions are provided as defaults by the software. The activation functions were automatically selected by the software from the set of five generics: linear, hyperbolic tangent, logistic, exponential and sinus. The learning process was provided according to BFGS algorithm (Broyden-Fletcher-Goldfarb-Shanno). The maximum number of iterations was 2000 with residual squared error not greater than 10^{-7} in the window of 20 epochs.

Three core approximations were identified (Fig.3). Apparently, the approximation 1-2-1 appears to be the best fitted with the smallest maximum error. This neural network was selected to the jackknife perturbation procedure.

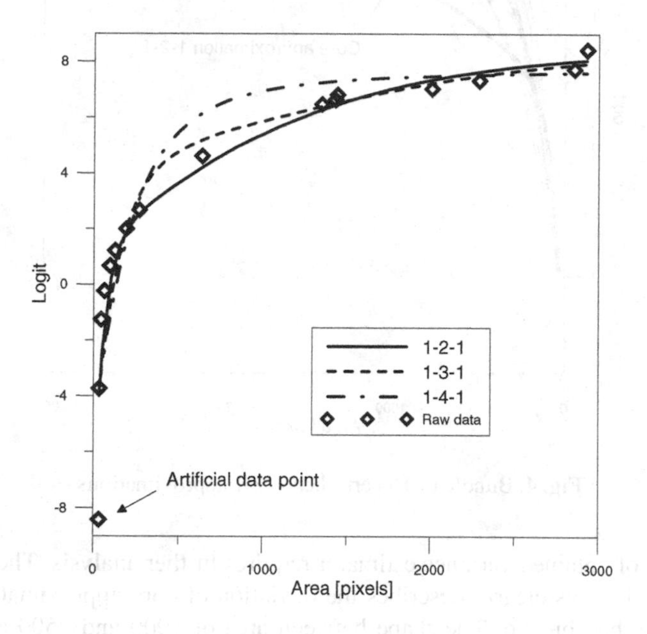

Fig. 3. Core approximations 1-2-1, 1-3-1 and 1-4-1

2.3 Jackknife Estimation of Uncertainty

Next, the structure of 1-2-1 neural network approximator was fixed i.e. the topology and activation functions were fixed while the weights and biases remain random. The jackknife procedure was applied to the data set and sequentially delete-1 approach lead to a bundle of 16 perturbed approximations (Fig.4).

It is shown that the bundle of functions does not lay symmetrically around the core approximation. Formally, the residual deviations (perturbed vs. core) at raw data points (learning nodes) satisfy the requirement of symmetry as a whole but it does not satisfied at the particular node. This asymmetry property requires further, more detailed investigation however there is a reasonable suspicion that such effect is ANN's analogue of the well-known Runge's phenomenon.

The bundle of obtained approximations was processed according to jackknife approach: from (eq.7) to (eq.13). The evaluated jackknife variance estimator (Fig.5) appear to be a non-monotonic function.

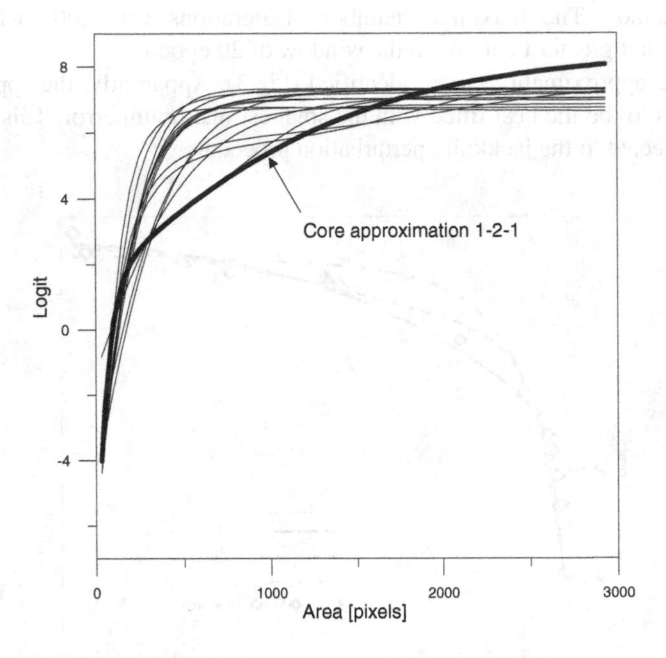

Fig. 4. Bundle of 16 perturbed 1-2-1 approximations

The shape of obtained variance estimator requires further analysis. The peak shown at approx. 800 pixels clearly describes the deviation of core approximation from the mean of perturbed bundle. The shape between area of 1000 and 2500 appears to be more strange. The core approximation goes approximately through the perturbed bundle but the minimum of variance is achieved at the right bound of area.

The specific problem that requires a separate consideration is accuracy of repeatability of STATISTICA Automatic Neural Networks module and – perhaps – averaging of identified neural networks.

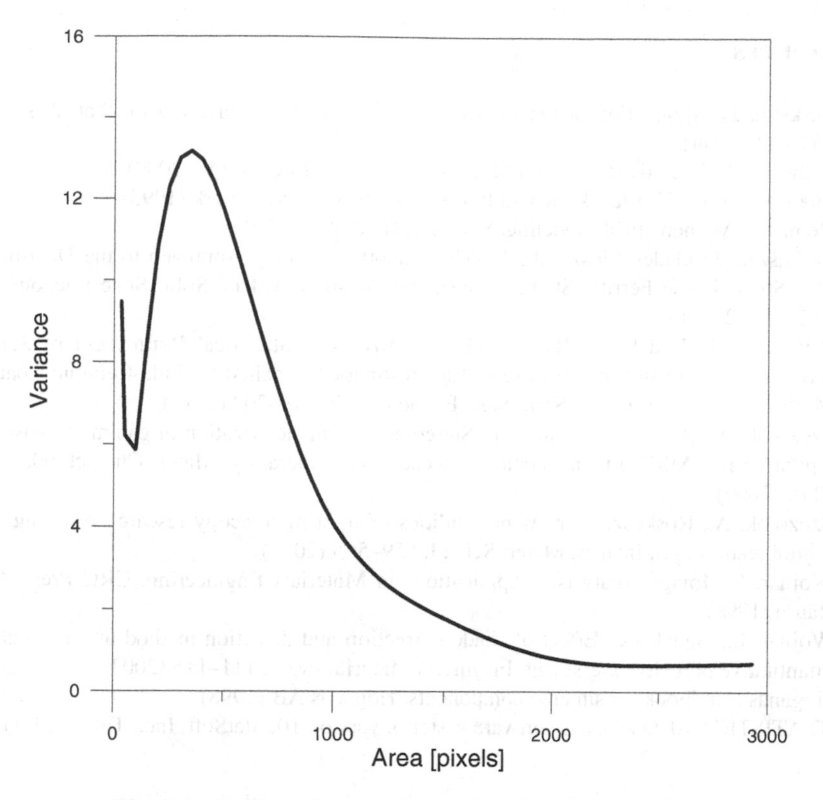

Fig. 5. Jackknife variance estimator for 1-2-1 approximation

3 Conclusions

At the beginning of conclusions it should be noted that the primary goal: to confirm the basic applicability of the jackknife method to neural network approximators has been achieved and obtained results are meaningful, however, they look slightly strangely and temporarily they may be treated only qualitatively.

Firstly, it is clear that further investigation should concentrate on uniqueness and repeatability of neural network approximations obtained from Statistica Automatic Neural Network module. An important clue is to check that the output means demonstrate uniform convergence for the sequence of fixed-topology neural network approximators.

Secondly, the set of activation functions provided by Statistica should be intensively tested on different datasets to select the most appropriate topology, activation function and other parameters of learning process to obtain the best fitted model and further reliable confidence bands obtained from jackknife method. It is important because a typical materials science engineer is rather unfamiliar with the advanced computer image analysis, the neural network regression or the jackknife method and the such group of users requires the completely automated data processing without any manual intervention.

References

1. Berkson, J.: Application of the Logistic Function to Bio-Assay. J. Am. Stat. Assoc. 39, 357–365 (1944)
2. Hilber, J.M.: Logistic Regression Models. CRC Press, Boca Raton (2009)
3. Shao, J., Tu, D.: The Jackknife and Bootstrap. Springer, New York (1995)
4. Heinz, S.: Mathematical Modeling. Springer, Heidelberg (2011)
5. Pietraszek, J., Gądek-Moszczak, A.: The Smooth Bootstrap Approach to the Distribution of a Shape in the Ferritic Stainless Steel AISI 434L Powders. Solid State Phenom. 197, 162–167 (2013)
6. Pietraszek, J., Radek, N., Bartkowiak, K.: Advanced Statistical Refinement of Surface Layer's Discretization in the Case of Electro-Spark Deposited Carbide-Ceramic Coatings Modified by a Laser Beam. Solid State Phenom. 197, 198–202 (2013)
7. Szczotok, A., Richter, J., Cwajna, J.: Stereological characterization of gamma ' phase precipitation in CMSX-6 monocrystalline nickel-base superalloy. Mater. Charact. 60, 1114–1119 (2009)
8. Szczotok, A., Roskosz, S.: New possibilities of light microscopy research resulting from digital recording of images. Mater. Sci. 23, 559–565 (2005)
9. Wojnar, L.: Image Analysis – Applications in Materials Engineering. CRC Press, Boca Raton (1998)
10. Wojnar, L., Gądek, A.: Effect of shade correction and detection method on the results of quantitative porosity assessment. Inżynieria Materiałowa 3, 111–116 (2003)
11. Höganäs Handbook for sintered components. Höganäs AB (1998)
12. STATISTICA (data analysis software system), version 10. StatSoft, Inc., Tulsa (2011)

Algorithms for Solving Frequency Assignment Problem in Wireless Networks

Radosław Józefowicz, Iwona Poźniak-Koszałka, Leszek Koszałka, and Andrzej Kasprzak

Department of Systems and Computer Networks,
Wroclaw University of Technology,
Wroclaw, Poland
{radoslaw.jozefowicz,iwona.pozniak-koszalka,leszek.koszalka, andrzej.kasprzak}@pwr.wroc.pl

Abstract. In this paper, the fixed frequency or channel assignment problem, so-called FAP/CAP, is considered. The problem is general but it can be taken as relevant for real-world cellular GSM/UMTS networks. There are presented five algorithms to solve this problem. Two of them are created by the authors of this paper. These algorithms: *RHA* and *RKA* have been tested, evaluated and compared with previously proposed assignment algorithms on real-life examples. The obtained results showed that the proposed algorithms can provide efficient assignment to respecting cells requirements.

Keywords: wireless network, channel assignment, algorithm, experimentation system, simulation.

1 Introduction and Motivation

Nowadays Frequency Assignment Problem (*FAP*) is getting more and more popularity because of still increasing demand on the radio frequency spectrum that is limited by natural resource. The radio frequency spectrum can be divided into discrete channels that can be assigned to a particular transmitter. One common example of cellular mobile phone network is GSM / UMTS. Because of increasing prices for the radio spectrum it is crucial for the providers to try to assign channel/frequencies as efficient and effectively as possible to reduce overall costs [1].

Any given radio spectrum can be split into a set of radio channels that can be assigned to transmitters only when there is no interference between frequencies. *FAP* can be divided into three categories [2], namely, Fixed Channel Assignment (*FCA*), Dynamic Channel Assignment (*DCA*), and Hybrid Channel Assignment (*HCA*). *FCA* is the most popular schema of this problem. The area is split into cells which require number of assigned channels. In this model the demand on required channels in each cell is constant and does not change in time domain. The other distinguished schemas are: (i) *DCA* in which required number of channels in cells is not constant and varies over time domain, however, such a model is less efficient than *FCA* model under high

A. Bădică et al. (eds.), *Recent Developments in Computational Collective Intelligence*, Studies in Computational Intelligence 513,
DOI: 10.1007/978-3-319-01787-7_13,

load conditions [3]; (ii) *HCA* being a combination of *FCA* and *DCA*., where a number of channels is assigned statically to cells and the rest is assigned dynamically due to the current traffic in the network. In this paper, *FCA* schema is taken into consideration.

The considered *FAP* problem can be also split into four categories based on optimization function, the given requirements and given constraints which must be respected [4]. There are: (i) Maximum service (*max-FAP/CAP*) where the objective is to assign as many channels as possible to transmitters in the network. This scenario can be used in case when there is no solution for considered problem; (ii) Minimum order (*mo-FAP/CAP*) where number of used frequencies in the network should be minimized (to respecting all constraints and obtaining required quality of service); (iii) Minimum span (*ms-FAP/CAP*) where the objective is to find an assignment of frequencies to transmitters that minimizes the span with the difference between the minimum and the maximum frequencies given; (iv) Minimum interference (*mi-FAP/CAP*) where the penalty array is applied to describe the cost function, and the aim is to minimize a sum of the penalties incurred by channels selection. In this paper the maximum service (*max-FAP/CAP*) is considered with a few modifications – our model considers just co-channel interference. The algorithms designed and implemented by the authors of this paper can be used successfully in minimum span (*ms-FAP/CAP*) problem as well.

Summarizing, the objective of the paper is to find non-conflict channel assignment to respecting cells requirements of frequencies needed to serve users.

The rest of the paper is organised as follows: Section 2 contains the used nomenclature, problem statement and quick review of the problem complexity. In Section 3, the five implemented algorithms are briefly described. Section 4 presents the experimentation systems. Section 5 contains the results of investigations that are discussed as well. Finally, in Section 6, the conclusion appears.

2 Problem Statement

FAP is NP-hard class problem, so it means there is no polynomial time algorithm to solve this problem [5]. It has already been proven that *FAP* is equivalent to generalized graph-colouring problem, which is precisely NP-hard [4]. Thus, there is proposed number of heuristic algorithms that could deal with this problem efficiently [6], [7], and [8].

In the *FAP/CAP* problem, a given radio spectrum is divided into N channels between the minimum f_{min} and the maximum f_{max} frequencies, so-called span expressed by (1), where Δ denotes radio channel width.

$$N = \frac{f_{min} + f_{max}}{\Delta} \quad (1)$$

The most popular model that describes considered network is that using undirected graph $G=(V, E)$. In such a model each transmitter is considered as a node $v \in V$, and edges from v to w denoted as $\{v, w\} \in E$ represent interference between transmitters.

Thus that representation of the *FAP/CAP* problem is strongly connected with graph-colouring theory [9]. Each node in given set of nodes has Cartesian coordinates in two-dimensional plane known in advance. The distance *d* between points is calculated as Euclidean distance. Each node (transmitter) is described by two ranges – transmission range T_{tr} and interference range T_{ir}. If transmission range of one node T_{tr} $_{collides}$ with interference range T_{ir} of another node then these nodes cannot have assigned the same frequency – there is an edge between considered vertexes in the graph model (2).

$$Edge\ e = \{v, w\} \in E,\ d < T_{tr}(v) + T_{ir}(w) \quad (2)$$

In Fig.1 an interpretation of the introduced notions is illustrated.

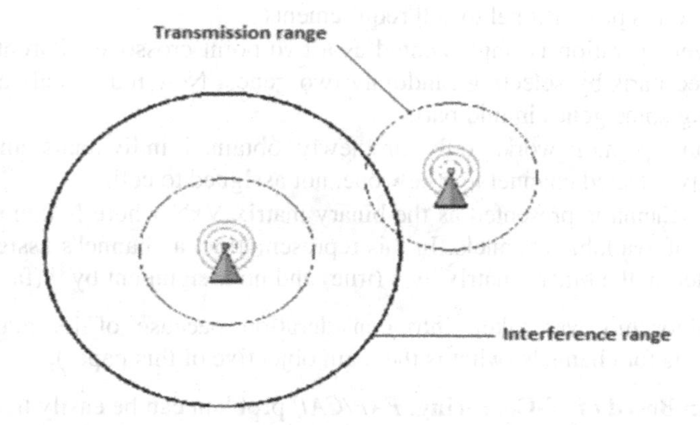

Fig. 1. Transmission and interference range of transmitter

The modified maximum service *FAP/CAP* problem is considered. The model is formulated by expressions (3) – (6), where *m(v)* denote the demand of channels in *v* and *n(v)* denote number of channels assigned to *v* [2] with constraint $n(v) \leq m(v)$.

$$max \sum_{v \in V} n(v) \quad (3)$$

$$x_{vf} + x_{wf} \leq 1 \quad (4)$$

$$\forall \{v, w\} \in E, f \in F(v) \cup f \in F(w) \quad (5)$$

$$f_{min} \leq f \leq f_{max} \qquad \forall f \in F \quad (6)$$

Variable x_{vf} is binary representation, $x_{vf} = \{0,1\}$, i.e., if channel is assigned to *v* then $x_{vf} = 1$; otherwise $x_{vf} = 0$. The maximum and the minimum frequencies are defined as f_{min} and f_{max}.

Such a representation of the problem guarantees maximum service (*max-FAP/CAP*) type of *FAP/CAP*. Equation (3) ensures that the number of assigned channels is maximized and at the same time there is no waste of channels assigned to cells – each cell obtains (if it is possible) just required number of channels.

3 Assignment Algorithms

3.1 Algorithms Based on Known Ideas

Ngo-Li Generic Genetic Algorithm (modified). The detailed description of the classical version of the algorithm can be found in [8]. For the purposes of this paper we designed and implemented a modification of Ngo-Li algorithm. A modified version is based on the following concepts:

- Initial population is generated randomly; however, the number of selected channels in each cell is proportional to cell requirements.
- Crossover operation is implemented as a two-point crossover. Parents are divided into three parts by selecting randomly two genes. New individuals are created by swapping some genes in mid part.
- Mutation operator works only on newly obtained individuals and it changes randomly selected channel to a new one, not assigned to cell.
- Coding schema is presented as the binary matrix VxN where N denotes maximum number of available channels. In this representation, a channel's assignment to cell is denoted in the binary matrix by 1 (true) and no assignment by 0 (false).

Ngo-Li algorithm was taken into consideration because of its maintaining cell requirements for channels (what is the main objective of this paper).

Algorithm Based on T-Colouring. *FAP/CAP* problem can be easily transformed into graph theory domain. For the purposes of this paper a modified greedy algorithm based on T-Colouring [1] was implemented and examined.

Random Assignment Algorithm (*RAA*). Many of complex optimization problems can be solved by using algorithms based on randomness [9]. Considering *FAP/CAP* the obvious solution is to assign a new channel to randomly selected cell. The only constraint is that the cells cannot interfere with others. The process of random assignment was proposed by the authors of this paper. A pseudo-code of the implemented algorithm is as follows:

RAA (Random Assignment Algorithm):

1. Assign to each cell channel using greedy algorithm
   k = maximum assigned channel number
   ch = available channels
   i = number of iterations
2. While $i! = ch$
   -assign to randomly selected cell new channel increment i

3.2 The Created Algorithms

Radek Hybrid Assignment (*RHA*). This algorithm is inspired by graph-colouring problem. A given cellular radio network is treated as an undirected graph. The idea of *RHA* is based on the following activities:

- In order to have in each cell at least one channel - one frequency per cell is assigned using greedy graph-colouring techniques.
- When each cell has one channel assigned a queue is built. The order in a queue is specified by fitness function defined by (7):

$$\boldsymbol{fitness = u - (n \cdot m - s)} \tag{7}$$

  where u –the number of users in cell, n – the number of already assigned channels, m – multiplication parameter, s - channels needed to signalisation.
- The proposed fitness function ensures that cells with higher requirements are considered as the first.
- After each assignment a new value of fitness function is evaluated and cell is reinserted to queue with a new priority.
- The algorithm works until built queue is not empty. Queue can be considered as empty only if one of the following conditions is matched: (i) No conflict-free assignment is possible; (ii) All cells have required number of channels assigned.

The objective is to use the lowest available channel number that allows conflict-free assignment. A pseudo code of the proposed procedure is as follows: (1) For each i vertex in V checks if i vertex is allowed and if yes return i. (2) If loop ends with no free channel available return -1.

RHA is a hybrid algorithm which provides satisfying results. However, the approach used in *RHA* strongly favours the cells with higher requirements, thus the initial assignment is necessary. A pseudo code of this algorithm is as follows:

***RHA* (Radek Hybrid Assignment):**
1. Assign to each cell a channel using greedy algorithm.
2. Build priority queue of cells based on number of users in each cell.
3. While priority queue is not empty
   - assign to first cell in priority queue new channel,
   - remove top element from priority queue,
   - decrease number of users in cell needed to be served,
   - reinsert cell to priority queue.

Radek Knapsack Assignment (*RKA*). This algorithm transfers *FAP/CAP* problem into simplified knapsack problem [10]. Each cell is considered as a separate knapsack which has to have number of channels inside. Because of fact that radio spectrum is limited resource the objective is to find the best optimal solution that means to provide as good coverage as it is possible. This algorithm tries to achieve best percentage coverage – cells with higher requirements are not favoured as much as in *RHA*. The same like in *RHA* cells are stored into queue but in this case with ascent priority. The priority is computed by the introduced fitness function (8):

$$\boldsymbol{fitness} = \frac{\boldsymbol{n} \cdot \boldsymbol{m} - \boldsymbol{s}}{\boldsymbol{u}} \quad \textbf{(8)}$$

where parameters are defined below (7). Such fitness function ensures that each cell is to be serving at least once and at the same time provides more fair assignments of channels between cells. The special parameter *beta* is in use to decide while assignment is about to stop (*fitness* $\leq$ ***beta***). The default value equal to 1 means only that the required number of channels is assigned to each cell. *Beta* parameter makes this algorithm very flexible – easily can provide assignment with redundancy or assignment with the lowest cost as it is possible. The idea of using queue and conditions of stopping the algorithm work is the same as in *RHA* (*beta* parameter is considered additionally). A pseudo-code of *RKA* algorithm is as follows:

***RKA* (Radek Knapsack Assignment):**

1. Assign to each cell f (fitness)=0
2. Set value of *beta* parameter (default is 1)
3. Build priority queue of cells based on fitness function in descent order
4. While priority queue is not empty
   - assign to top element new channel,
   - remove top element from priority queue,
   - evaluate new value of fitness function,
   - reinsert cell to priority queue

4 Experimentation System

For performing benchmark's tests there were two simulators used. The first one is simple console application written in C++ by the authors of this paper. This program contains the implementations of *RHA*, *RKA* and *RAA*. The second simulator is CHALLOC program developed by Ignaczak [11]. It allows for performing along with Generic Genetic Algorithm, and Algorithm based on T-Colouring. The experimentation system as input-output plant is shown in Fig. 2

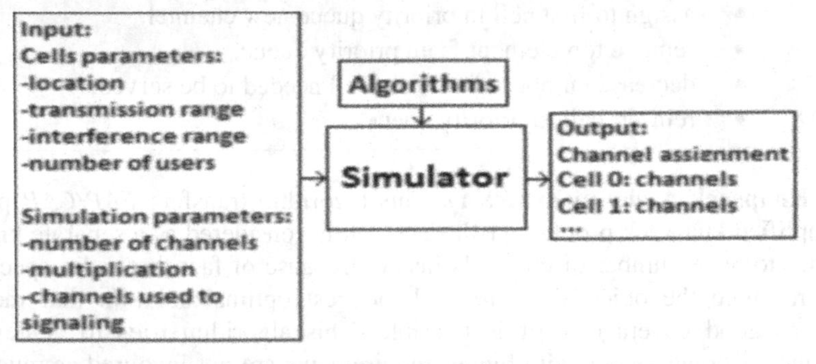

Fig. 2. The input-output experimentation plant

In order to achieve maximum efficiency *RKA* and *RHA* use so-called priority queue structure as a cells container. The main advantage of using this structure is very fast access (O(1) – complexity) to an element with the highest priority. The time needed to insert new element to queue is better than linear as well (O(log(n))). Thus it is obvious to choose this structure to contain cells. Two benchmark networks were used to perform tests. The first one was the P1 instance of *Philadelphia* instances (Fig. 3). The second one was the instance *siemens2* from COST259 project [6].

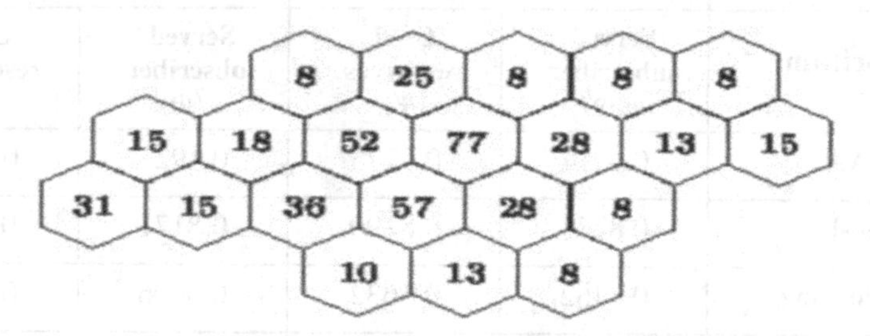

Fig. 3. *Philadelphia* P1 instance [6]

5 Investigation

The objective of the investigations was to compare efficiency of the considered algorithms in the same conditions – using the same values of parameters: *N, m, s*, where *N* – number of available frequencies, *m* – multiplication, *s* – channels used for signalisation. For each given network there were fixed location parameters, ranges (transmission and interference) and requirements which did not change during tests.

RAA and Ngo-Li Generic Genetic algorithms were repeated ten times to ensure reliability (both algorithms are probabilistic).

Measures of Efficiency. Three metrics were proposed to compare the five algorithms examined. These metrics are strongly connected to cellular GSM/UMTS networks.

Percentage of served subscribers (Φ). This metric is designed to describe what percentage of users is served in considered networks. The metric is calculated by (9).

$$\phi = \frac{\sum_i \min\left([\max(n_i \cdot m - s; 0)]; a_i\right)}{\sum_i a_i} \tag{9}$$

Percentage of used resources (Ψ). This metric reflects the tendency of operators to use as few channels as possible. The metric is computed by (10).

$$\Psi = \frac{\sum_i \min\left(n_i \cdot m; a_i + s\right)}{\sum_i a_i} \tag{10}$$

Experiment Design. The considered networks parameters were taken as follows - for Philadelphia P1: N=40, m=8, s=1; for siemens2: N=76, m=1, s=0. The internal parameters for Ngo-Li Genetic algorithm were: Number of generations=100000, Number of individuals in the population=50, Minimal change between generation=~ 1%, Crossover probability=95%, Mutation probability=5%. *Beta* parameter in *RKA* was set to 1. The number of iterations for RAA was set to 50 for *Philadelphia* P1 and to 100 for *siemens2*.

Results. The obtained results are presented in Table 1 and in Fig. 4 – Fig. 7, where *Percentage* means results of calculations with (9) and (10).

Table 1. Results for *Philadelphia* P1 and for *siemens2* instances

Algorithm	*Philadelphia* P1 instance		*siemens2* instance	
	Served subscribers (Φ)	Used resources (Ψ)	Served subscribers (Φ)	Used resources (Ψ)
RAA	0.6914	0.6225	0.1921	0.9837
Ngo-Li	0.8777	0.8190	0.8171	0.9911
T-Colouring	0.8462	0.5632	0.3406	0.9939
RHA	0.8711	0.7971	0.7123	0.9384
RKA	0.9667	0.8933	0.7258	1.0000

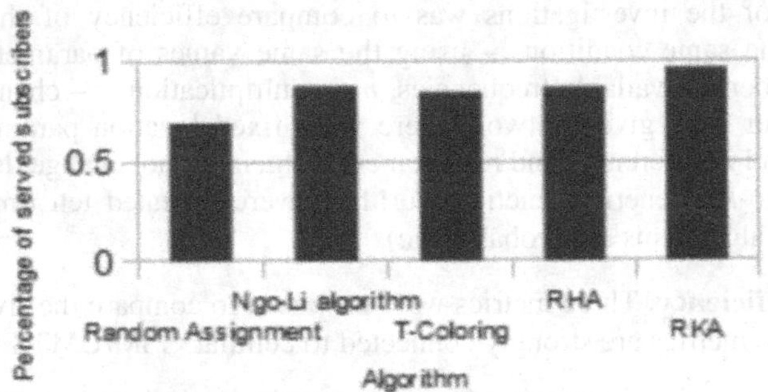

Fig. 4. Percentage of served subscribers for *Philadelphia* P1 instance

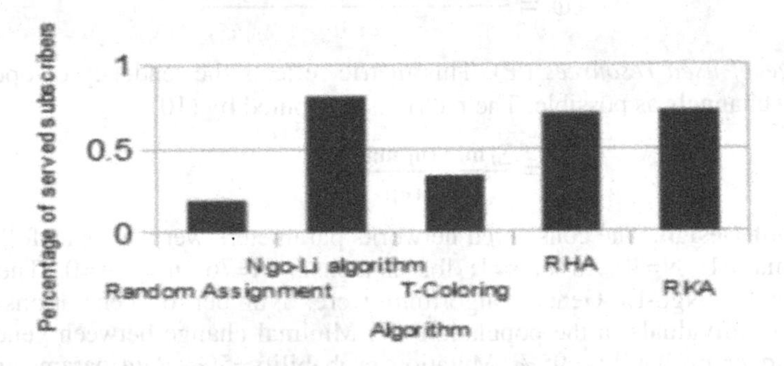

Fig. 5. Percentage of served subscribers for siemens2 instance

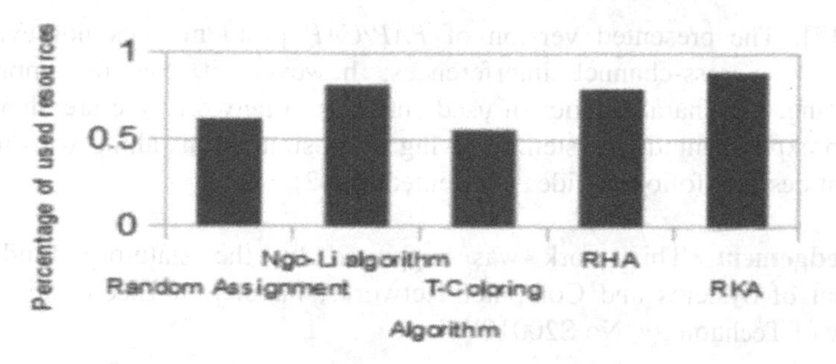

Fig. 6. Percentage of used resources for Philadelphia P1 instance

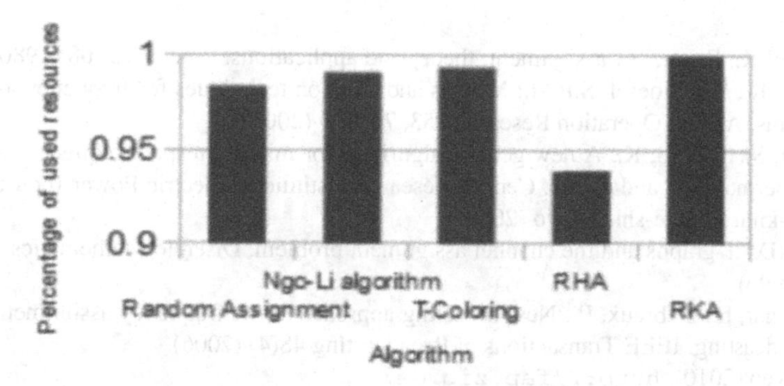

Fig. 7. Percentage of used resources for siemens2 instance

It may be observed that the newly proposed assignment algorithms perform well (an exception is metric - used resources - for siemens2). For both tested networks, as expected, the worst results were obtained by RAA. In simple exemplary instance P1 the proposed algorithm RKA achieved the best results for all considered metrics. Considering a difficult network like siemens2 instance the best result was obtained by Ngo Li algorithm, however, it was only slightly better than RKA.

6 Conclusion

Two new algorithms to solve maximum service *FAP/CAP* problem were presented and evaluated. Using these algorithms we obtained very promising results for *Philadelphia* P1 network and for the difficult *siemens2* instance. RKA is more flexible because of introducing so-called *beta* parameter that is responsible for percentage of used resources in assignment process. *RKA* and *RHA* are designed to implement different fitness functions that allow taking into account more parameters like weight of the users in cells or cell assignment cost.

We expect that our approach can be successfully used in Dynamic Channel Assignment as well. The *RHA* algorithm can be improved by changing the technique for initial assignment, e.g., using Ant-Colony optimization for graph-colouring problem [7]. The presented version of *FAP/CAP* problem does not evaluate all constraints, as cross-channel interferences; however, it can be improved by implementing, e.g. characteristics of used antennas in network. We are also planning to develop experimentation system allowing more simulations along with multistage experiment designs, following ideas presented in [12].

Acknowledgement. This work was supported by the statutory funds of the Department of Systems and Computer Networks, Faculty of Electronics, Wroclaw University of Technology, No S20010W4.

References

1. Hale, W.K.: Frequency assignment: theory and applications. Proc. IEEE 68 (1980)
2. Aardal, K., Van Hoesel, S.P.M.: Models and solution techniques for frequency assignment problems. Annual Operation Research 153, 79–129 (2007)
3. Matsui, S., Tokoro, K.: A new genetic algorithm for minimum span frequency assignment using permutation and clique. Central Research Institute of Electric Power Industry 2-11-1 Iwado-kita, Komae-shi, Tokyo (2000)
4. Liu, D.D.: T-graphs and the channel assignment problem. Discrete Mathematics 161, 197–205 (1996)
5. Idounghar, L., Debreux, P.: New modeling approach to the frequency assignment problem in broadcasting. IEEE Transactions of Broadcasting 48(4) (2006)
6. FAP Web (2010), `http://fap.zib.de/`
7. Salaril, E., Eshghi, K.: An ACO algorithm for graph coloring problem. Int. J. Contemp. Math. Sciences 3(6), 293–304 (2008)
8. Ngo, Y.C., Li, V.O.K.: Fixed channel assignment in cellular radio networks using a modified genetic algorithm. IEEE Trans. Veh. Technol. 47 (1998)
9. Baba, N.: Convergence of a random optimization method for constrained optimization problems. Journal of Optimization Theory and Applications 33(4) (1981)
10. Wroblewski, P.: Algorithms, Data Structures, Programming Techniques. Helion (2003) ISBN: 8373611010 / 83-7361-101-0
11. Ignaczak, J.F.: Computer decision making system using evolutionary algorithms for solving radio-channel splitting to cells in wireless networks. MSc Project, Faculty of Electronics, Wroclaw University of Technology, Wroclaw (2011)
12. Koszalka, L., Lisowski, D., Pozniak-Koszalka, I.: Comparison of allocation algorithms for mesh structured networks with using multistage simulation. In: Gavrilova, M.L., Gervasi, O., Kumar, V., Tan, C.J.K., Taniar, D., Laganá, A., Mun, Y., Choo, H. (eds.) ICCSA 2006. LNCS, vol. 3984, pp. 58–67. Springer, Heidelberg (2006)

Comparing Smoothing Technique Efficiency in Small Time Series Datasets after a Structural Break in Mean

Emil Scarlat[1], Daniela Zapodeanu[2], and Cociuba Mihail Ioan[2]

[1] Academy of Economic Studies,
Faculty of Economic Cybernetics, Statistics and Informatics
[2] University of Oradea, Faculty of Economics
danizapodeanu@yahoo.com, cociuba@gmail.com

Abstract. Using daily RON/EURO exchange rate data for the 01:2005 to 03:2013 period we test the presence of structural break using the Zivot-Andrews test and PELT algorithm. After the identification of structural breaks we generate small time series consisting of 10, 30 observations starting from the moment of the break and apply the following smoothing techniques: simple moving average, exponential moving average, a Grey model GM(1,1). We identify the best smoothing techniques using sum of squared errors (SSE) and mean relative error (MRE), in small samples GM(1,1) over-performed the moving average and the exponential moving average smoothing techniques.

Keywords: grey models, smoothing, exchange rate, small sample.

1 Introduction

Prediction and smoothing in time series is a challenging task due to the dynamic and special character of time series. Financial time series, in particular, have specific characteristic: long memory [Granger & Ding, 1995] and follow heteroskedastic and leptokurtic distributions. Grey models were developed by [Julong, 1982] who found that in the uncertain models the information is incomplete, the different level of available information in a system generates the following classification [Liu & Forrest, 2010] of the systems: - completely known information: white systems, - completely unknown information: black systems, - partially known information and partially unknown information: grey models.

[Tseng, Yu, & Tzeng, 2001] developed a hybrid GM(1,1) model with a moving average component that can be used in forecasting time-series with a seasonality component, the hybrid GM(1,1) model is tested against a SARIMA model using the following criterion: mean squares of errors (MSE), mean absolute error (MAE), mean absolute percentage error (MAPE); the study finds that the hybrid GM models has the lowest results for the MSE, MAE, MAPE. [Hung, Huang, Lin, & Hou, 2010] compares the forecasting power of the moving average and grey models with a hybrid suport vector regression (SVN) with a grey

Studies in Computational Intelligence 513,
DOI: 10.1007/978-3-319-01787-7_14,

component, they find that the hybrid models performance better than moving average or the grey models.

[Ho, 2005] shows the use of grey models through the Grey Relation Analysis on the banks performance on a small sample. The grey models can be used in improving forecasting models of the traditional statistics methods, [Bianco, Manca, Nardini, & Minea, 2010] applies a trigonometric grey model with rolling mechanism on the electricity consumption in Romania and a Holt&Winters exponential smoothing method, with the two models having similar results. The grey models are used to model the financial series in small sample [Tseng et al., 2001], also non-stationary data forecast is improved using grey models [Yang, Chen, & Lu, 2012]. [Wang, Lin, Huang, & Wu, 2012] applies an GM -GARCH model on the option market in order to eliminate heteroskedasticity. By comparing traditional smoothing techniques with Grey models smoothing on time-series dataset we analyses the smoothing efficiency on small datasets.

The remaining of the article is organized as follows: Section 2 outlines the methodology; Section 3 describes the dataset, presents the unit-root test and structural break analysis ; Section 4 presents the results of smoothing techniques; Section 5 concludes.

2 Methodology

If we have a time series a_t with t=1..n then it can be defined as the sequence of raw data $X^{(0)} = (x^{(0)}(1), x^{(0)}(2), ..., x^{(0)}(n))$ where $x^{(0)}(t) = a_t$. Then the accumulation generated sequence is $X^{(1)} = (x^{(1)}(1), x^{(1)}(2), ..., x^{(1)}(n))$ where $x^{(1)}(n) = \sum_{i=1}^{n} a_i$.

The original form of grey model GM(1,1) is $x^{(0)}(k) + ax(1)(k) = b$ the whitenization equation of the GM(1,1) is $\frac{\partial x^{(1)}}{\partial t} + ax^{(1)} = b$. Solving the whitenization equation we obtain the time response function as $x^{(1)}(t) = (x^{(1)}(1) - \frac{b}{a}) * \mathrm{e}^{-at} + \frac{b}{a}$. The estimated values from the GM(1,1) models are given by

$$\hat{x}^{(0)}(t) = \hat{x}^{(1)}(t) - \hat{x}^{(1)}(t-1) = (1 - \mathrm{e}^{a}) * (x^{(0)}(1) - \frac{b}{a}) * \mathrm{e}^{-a(t-1)} \quad (1)$$

for t=1,2,.., n. The parameters of the GM(1,1) equation are estimated using the method of least squares. The moving average can be used in small samples due to its simple structure, the smoothed series is calculated as the mean of the last k observations:

$$MA_t = \frac{1}{k} \sum_{n=0}^{k-1} a_{t-n} = \frac{x_t + x_{t-1} + \ldots + x_{t-k+1}}{k} = MA_{t-1} + \frac{x_t - x_{t-k}}{k}. \quad (2)$$

The exponential moving average is defined as:

$$EMA_t = \alpha * a_{t-1} + (1 - \alpha) * EMA_{t-1} \quad (3)$$

where α represents the smoothing factor and can take value between 0 and 1 and the first term is $EMA_1 = a_1$.

In order to test the stability of the series the following unit-root test are used: augmented Dickey&Fuller test (ADF) and Phillips&Perron (PP) test which have the null hypothesis H_0 that the series is integrated of order 1, respectively I(1), and the alternative hypothesis H_1 that the series is stationary, respectively I(0), while the KPSS tests has null hypothesis H_0 that the series is stationary, I(0).

The ADF, PP, KPSS tests doesn't take into account any structural changes in the time-series, in which situation [Perron, 1989] it may lead to rejecting the series stationarity even if the series is stationary (with a break in the intercept, trend or both). In order to eliminate the possibility of rejecting the stationarity of series we apply the [Zivot & Andrews, 1992], who extended the Dickey &Fuller test by allowing for a break in intercept, trend and both. Because the unit root test power and size properties depends on the number of lagged terms used, in order to eliminate serial correlation we apply [Zivot & Andrews, 1992] methodology by allowing a flexible number of lagged terms (k), where k is determined by the t-statistics of the coefficient. [Sen, 2003] showed that the best model to use in testing the unit-root hypothesis for the Zivot-Andrews test is model C, with both a break in intercept and trend.

The detection of breakpoints in time series can be formulated as having the following null hypothesis [Killick, Fearnhead, & Eckley, 2011], H_0 the time series has no changepoint (m=0) and the alternative hypothesis H_1 where we have at least 1 changepoint (m>=1). We test for changepoints in the exchange rate using the following statistical criteria: penalized likelihood, quasi-likelihood and CUSUM; the breakpoint analysis is carried in the mean. The test statistics used in the PELT method implementation [Rebecca Killick & Eckley, 2012] has the following hypothesis:

H_0 : *no breakpoint* with the maximum log-likelihood $ML = \log p(y_{1:n}|\theta)$
H_1 : *one breakpoint* τ_1, where $\tau_1 \in (1, 2, \ldots, n-1)$ is the possible breakpoint date and the maximum log-likelihood $ML = \log p(y_{1:\tau_1}|\theta_1) + (y_{t:\tau_1}|\theta_2)$

By rejecting the null hypothesis H_0, a breakpoint is detected. From the moment of the breakpoint from the full sample we generate sub-samples of 10 and 30 observation and test the efficiency of smoothing techniques.

In order to compare the smoothing techniques efficiency we use the mean following indicators: mean relative errors (MRE) and the sum of squared errors (SSE):

$$MRE = \frac{1}{n} \sum_{i=1}^{n} \frac{|error|}{\hat{y}} \tag{4}$$

$$SSE = \sum_{i=1}^{n} (y_i - \hat{y}_i)^2 \tag{5}$$

3 Data Analyzes

The dataset consists of daily exchange rate between RON/EURO from January 2005 to March 2013, the dataset is obtained from the National Bank of Romania, www.bnr.ro

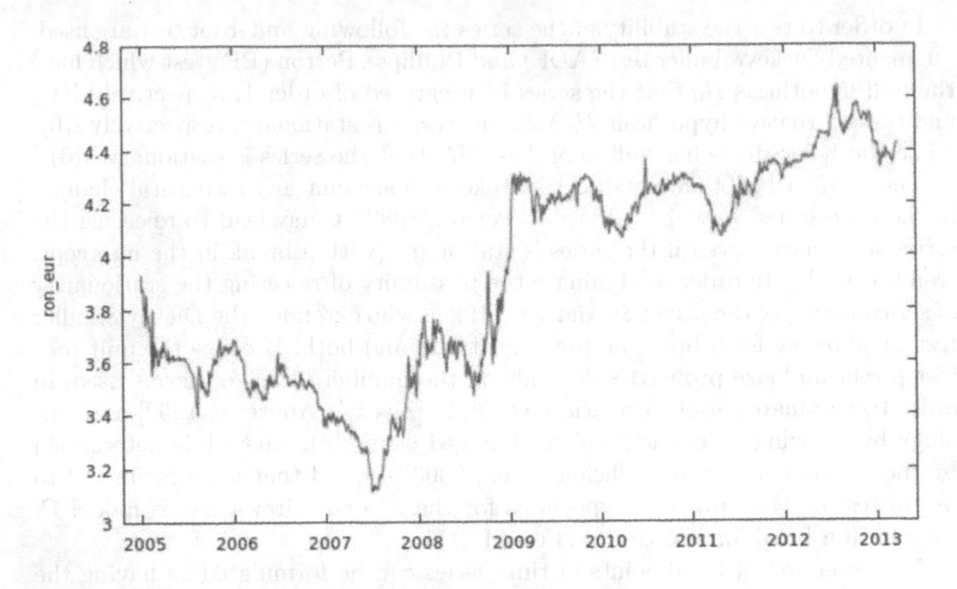

Fig. 1. Exchange rate evolution RON/EURO 2005-2012

In the analysed period for the RON/euro exchange rate we can identify two different trends, Fig.1, until august 2007 when the RON has appreciate in relation with the euro, while after august 2007 the RON is significantly depreciate. The exchange rate fluctuated between a minimum of 3.11 and a maximum of 4.64.

Table 1. Summary statistics

Mean	Median	Minimum	Maximum	SD	SK
3.9259	4.0908	3.1112	4.6481	0.4008	-0.1415
KT	**JB**	**Q(5)**	**Q(10)**	Q^2**(5)**	Q^2**(10)**
1.5540	189.32	10404	20708	2090	20706

Where SD, SK, KT, and JB denote standard deviation, skewness, kurtosis, and Jarque-Bera statistic, respectively. The Ljung–Box statistics, Q and Q^2 stat checks for serial correlation of inflation and squared inflation up to the 5th (10) order, the critical value for the Q(6), Q(12) respectively 16.81, (26.21), at 1% significance level. The critical value for the Jarque-Bera (JB) test is 5.991 at 5% significance level.

From the descriptive statistics, presented in Table 1, of the exchange rate, for the analyzed period, 2005:01 to 2012:11, the highest value was 4.64 in August 2012, while the lowest value was 3.11 in July 2007. The average value for the analysed period is 3.92. The exchange rate volatility, which can be measured through the standard deviation is 0.40. Also the series is negatively skewed; the

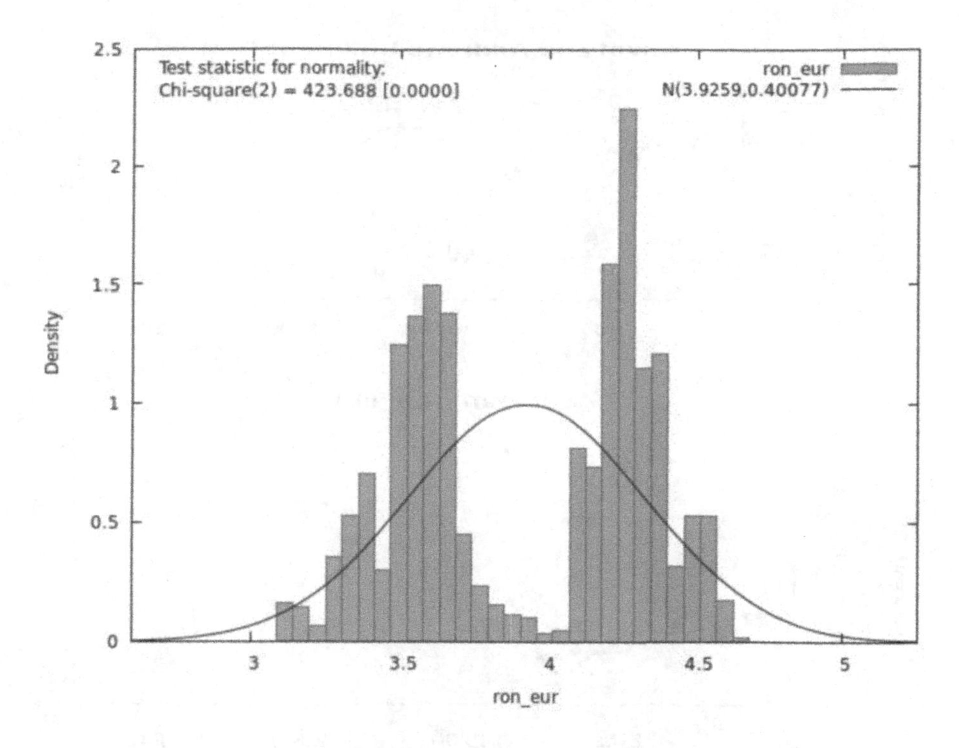

Fig. 2. Normality test RON/EURO

kurtosis value is below 3, which is the value for the normal distribution.The Jarque-Bera, Fig.2, test indicates that the distribution of exchange rate is non-normal, also the Q-statistics indicates serial correlation of exchange rate.

Table 2. Unit Root/stationarity test

ADF test	PP test	KPSS test
-3.430	-3.345	0.280

MacKinnons 5% critical value is 3.46 for the ADF and PP tests, the critical value for the KPSS test is 0.217 at 1% significance level, * denote significance at 1% levels.

Table 2 present the result for the unit root test, where ADF and PP test has the null hypothesis that the series is integrated of order 1, while KPSS null hypothesis is that the series is stationary, and ZA test allows for a break in intercept, trend or both. The number of lags (k) in ADF test was chosen using [Schwert, 1989] recommendation by setting k = 12*(T/100)0.25 and testing down the significance of the k lag coefficient; the number of extra regressors (k)

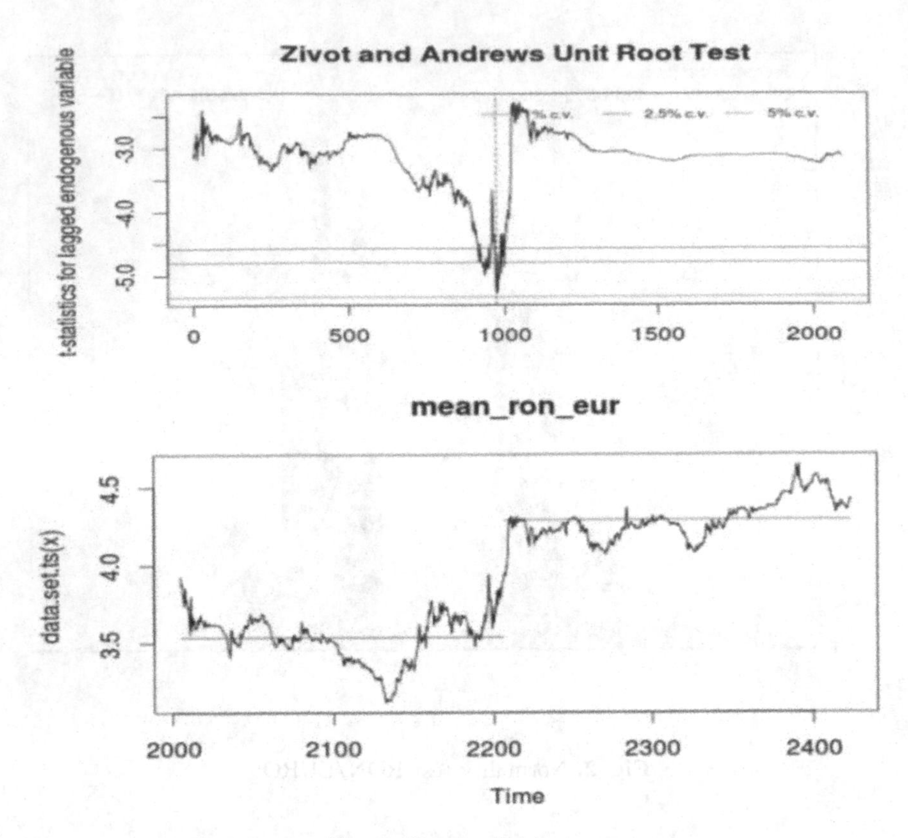

Fig. 3. Breakpoint tests

used in the Zivot-Andrews test are selected by allowing a flexible number of lagged terms (k) in order to eliminate serial correlation, where k is determined by the t-statistics of the coefficient. We find that we cannot reject the unit-root hypothesis following the ADF test at the 5% levels but we can reject the unit-root hypothesis at 10% significance, PP test also finds that the series has a unit-root at 5% significance level, the KPSS test rejects the stationarity of exchange rate for the analyzed period.

Based on the contradictory results from the ADF, PP and KPSS we apply the Zivot-Andrews test. The model used in the ZA test allows for a break both in the intercept and the trend, we reject the null hypothesis and conclude that the exchange rate is stationary with at least one breakpoint in the series mean.

The Zivot-Andrews test statistics is -5.29 with the number of extra regressors (k) used in the Zivot-Andrews test being selected based on their t statistics, the critical value is -5.34 at 1% significance level, -4.8 at 5% and -4.58 at 10% significance level.

From the Zivot-Andrews unit root test we find that the series is stationary at the 95% significance level, the structural break is identified at the 971

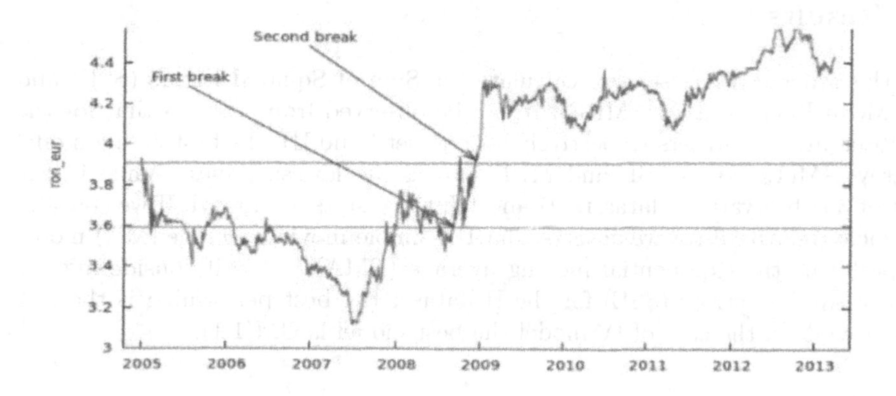

Fig. 4. Breakpoint date

Table 3. Smoothing performance

sum of squared errors (SSE)

Dataset	Period	Moving Average	Exponential MA	GM(1,1)
I	22.10.2008–04.11.2008	0.009272	0.018257	**0.007870**
II	22.10.2008–03.12.2008	**0.026563**	0.054944	0.039033
III	11.12.2008–24.12.2008	0.001159	0.002772	**0.001093**
IV	11.12.2008–27.01.2009	0.019634	0.132071	**0.089961**

mean relative error (MRE)

Dataset	Period	Moving Average	Exponential MA	GM(1,1)
I	22.10.2008–04.11.2008	0.82%	1.05%	**0.72%**
II	22.10.2008–03.12.2008	**0.68%**	0.93%	0.85%
III	11.12.2008–24.12.2008	0.26%	0.40%	**0.26%**
IV	11.12.2008–27.01.2009	**0.48%**	1.27%	0.26%

observations, respectively in 22.10.2008; the changepoint analysis identifies a breakpoint at the 1006 observation, respectively in 11.12.2008.

The following small dataset are generated from the series full sample:
- from the first breakpoint (2008-10-22): I series of 10 observations from 2008-10-22 until 2008-11-04, II series of 30 observations from 2008-10-22 until 2008-12-03
- from the second breakpoint (2008-12-11), III series of 10 observations from 2008-12-11 until 2008-12-24, IV series of 30 observations from 2008-12-11 until 2008-12-24.

We will test the following smoothing techniques: simple moving average, exponential moving average, a Grey model GM(1,1) on the generated datasets.

4 Results

On the generated datasets we calculate the Sum of Squared Errors (SSE) and the Mean Relative Error (MRE). It can be observed from Table 3 that for the 10 observations datasets respectively for dataset I and III, the best fitting model is grey GM(1,1) with SSE and MRE having the lowest values. While in the case of 30 observations datasets, II and IV, the results are mixed. If we consider the mean relative error we observe that the simple moving average (MA) model outperforms the exponential moving average (EMA), while if considering the sum of squared errors (SSE) for the II dataset the best performing is the MA model while in the case of IV model the best model is GM(1,1).

5 Conclusion

Analyzing the RON/EURO exchange rate data for the 01:2005 to 03:2013 period we find that the series has two structural breaks at the 971 observations, respectively in 22.10.2008; while the changepoint analysis identifies a breakpoint at the 1006 observation, respectively in 11.12.2008. We identify the best smoothing techniques using sum of squared errors (SSE) and mean relative error (MRE), the result being that in small samples the GM(1,1) over-performed the moving average and the exponential moving average smoothing techniques, while if the number of observation in the samples are increasing the results are mixed, the use of Grey models in small samples improves the accuracy of smoothing technique specially when the number of observations is low.

References

1. Bianco, V., Manca, O., Nardini, S., Minea, A.: Analysis and forecasting of non-residential electricity consumption in Romania. Applied Energy 87(11), 3584–3590 (2010)
2. Box, G., Jenkins, G., Reinsel, G.: Time series analysis: forecasting and control, pp. 1013–1015. Wiley (2011)
3. Granger, C.W.J., Ding, Z.G.: Some properties of absolute return: An alternative measure of risk. Annales dEconomie et de Statistique, 67–91 (1995)
4. Ho, C.-T.: Measuring bank operations performance: an approach based on Grey Relation Analysis. Journal of the Operational Research Society 57(4), 337–349 (2005)
5. Hung, C., Huang, X.-Y., Lin, H.-K., Hou, Y.-H.: Integrated Time Series Forecasting approaches using moving average, grey prediction, support vector regression and bagging for NNGC. In: The 2010 International Joint Conference on Neural Networks (IJCNN), vol. 16 (2010)
6. Julong, D.: The grey control system. Journal of Huazhong University of Science and Technology 3(9), 18 (1982)
7. Killick, R., Fearnhead, P., Eckley, I.: Optimal detection of changepoints with a linear computational cost. Journal of the American Statistical Association (2012) (accepted)

8. Killick, R., Eckley, I.: Changepoint: An R package for changepoint analysis. R package version 0.6, pp. 1–18 (2012), http://www.lancs.ac.uk/killick/Pub/KillickEckley2011.pdf
9. Liu, S., Forrest, J.Y.L.: Grey systems: theory and applications, vol. 68. Springer (2012)
10. Perron, P.: The great crash, the oil price shock, and the unit root hypothesis. Econometrica: Journal of the Econometric Society 57(6), 1361–1401 (1989)
11. Sen, A.: On unit-root tests when the alternative is a trend-break stationary process. Journal of Business & Economic Statistics (573) (2003)
12. Tseng, F., Yu, H., Tzeng, G.: Applied Hybrid Grey Model to Forecast. Technological Forecasting and Social Change, 291–302 (2001)
13. Wang, C.-P., Lin, S.-H., Huang, H.-H., Wu, P.-C.: Using neural network for forecasting TXO price under di erent volatility models. Expert Systems with Applications 39(5), 5025–5032 (2012)
14. Yang, Y., Chen, W., Lu, H.: A Fuzzy-Grey Model for Non-stationary Time Series Prediction. Appl. Math. (2) (2012)
15. Zivot, E., Andrews, D.: Further evidence on the great crash, the oil-price shock, and the unit-root hypothesis. Journal of Business & Economic Statistics 10(3), 251–270 (1992)

Part III
Language and Knowledge Processing Systems

Enhancing the Effectiveness of the Spelling Checker Approach for Language Identification

Nicholas Akosu and Ali Selamat

Software Engineering Department, Faculty of Computer Science and Information Systems, Universiti Tecknologi Malaysia
nickakosu@yahoo.com, aselamat@utm.my

Abstract. The ability to automatically identify the language of a text is often necessary for information retrieval in a multilingual environment such as the Internet. A general solution to this problem cannot be contemplated as long as most of the world's languages have not been subjected to a language identification study. The main objective of this study is to propose a new algorithm for improvement of the Spelling Checker technique which lends itself easily to language identification of under-resourced languages. The second objective is to study the effect of vocabulary expansion as a viable technique for enhancing the performance of language identification using the Spelling Checker approach. Experimental results show that this approach is a strong support for the Spelling Checker method and is capable of contributing to the development of digital resources for under resourced languages.

Keywords: Language identification, Under-resourced languages, Vocabulary extension, Spelling Checker.

1 Introduction

Research in the area of language identification (LID) has been going on for some time now and has reached very advanced stages with respect to the popular languages. Already there are many commercially available language identification packages capable of identifying several languages for example SILC [1] can handle language identification for over 100 languages and TextCat [2] can identify over 60 languages. However, these packages target the languages spoken by many people and not the minority languages i.e. those spoken by a few. One of the reasons for this state of affairs is due to lack of digital resources in the so called minority languages. However, this situation is already changing as more and more languages are now appearing on the Internet with some level of interesting content. Moreover, as the Internet is becoming more wide-spread there is need to create the capability of identifying a greater number of the world's languages. According to Ethnologue[3], more than 7000 languages are available on Earth. And yet less than 30% of these languages can be identified using available language identification techniques.

A. Bădică et al. (eds.), *Recent Developments in Computational Collective Intelligence*, Studies in Computational Intelligence 513,
DOI: 10.1007/978-3-319-01787-7_15,

Consequently, there is need to develop easy and effective techniques capable of being applied to many languages. Hence the Spelling Checker approach proposed by Piernaar [4] is seen by the authors as very promising. This technique has already been shown to be very accurate, giving 100% in identification of 11 official languages of South Africa. In addition it is possible to apply existing resources in the development of the language identification system and also to launch the system quickly even where Spelling Checkers do not exist for the given language [4]. In order to include such a language in the system all one needs to do is to assemble suitable documents in the required language and develop the required Spelling Checker using the wordlist-based method proposed by Prinsloo [5]. Thereafter the Spelling Checker is incorporated in the system thereby making it possible for such a language to be identified.

In this paper, we propose a new algorithm for vocabulary extension aimed at upgrading the existing Spelling Checkers for language identification. The algorithm yields two advantages namely, increasing the accuracy (lexical recall) of the Spelling Checker and secondly helping to provide a viable Spelling Checker that can be applied in other areas of NLP with respect to the particular language. This is especially useful in the areas of development of digital resources for the under-resourced languages which form the bulk of the World's languages.

The rest of the paper is structured as follows: Section 2 takes a look at related work in the area of language identification. In section 3 we present the proposed method while section 4 explains the experimental setup and section 5 presents results and conclusion.

2 Related Work

Many researchers have done useful work on language identification in recent times. Selamat [6] proposed an improved n-gram algorithm based on original n-gram and modified n-gram aimed at developing suitable methods for language identification of web documents. Issues of computational costs and scalability limitations were not addressed in that research. Chew et al. [7] also proposed another improved n-gram algorithm to validate the use of the Universal Declaration of Human Rights (UDHR) Act [8] as a training corpus for language identification. They investigated options for finding the optimal length for training data and improvement of language identification for closely related languages. Issues of under-resourced languages were not addressed. Brown [9] also used the N-gram approach for language identification of over 900 languages with impressive results. Vatanen [10] carried out a study of Naïve Bayes classifier and n-gram to investigate the handling of short text in language identification. They reported that accuracy for short text requires higher models and leads to slower execution. Piernaar [4] investigated the Spelling Checker technique for language identification of the 11 official languages of South Africa. Issues relating to extending the language identification method to other resource scarce languages were not addressed.

Yang [11] presented an n-gram-and-wikipedia joint approach aimed at overcoming the drawback of large scale training sets and linguistic knowledge that are required in existing language identification methods. The issues of under-resourced languages were not the focus of this research. Amine [12] presented a hybrid algorithm that combined k-means artificial ants and n-gram as alternative to supervised approaches. But the researchers did not address issues of application of this technique to other languages yet to be investigated. In another study, Yang, J., et al.[13] used the Maximum A Posteriori Linear Regression technique for language recognition and obtained impressive results. Rehurek [14] extended the dictionary method (character pentagram models) with a view to addressing limitations on language identification on very short text and handling text of multiple languages. They obtained slightly lower accuracy, though this was compensated by improved run time performance. Ramisch [15] investigated application of n-gram language models using the maximum likelihood estimator (MLE) to estimate probabilities in a training corpus. They applied the Good-Turing discounting with back-off smoothing and obtained very impressive results.

3 Methodology: Language Identification and Vocabulary Extension

The philosophy behind the effectiveness of this technique is based on the fact that a Spelling Checker with additional words will always be more capable of finding more words in a document that is written in that language. The principle for the extension is based on the same principle used in word processors whereby the user is always requested to decide if an unknown word is to be added to the dictionary. In this case the extension of the vocabulary (dictionary) concerns a group of words and is to be done as shown in figure 1.

The idea is to grow the language models by adding new words that are discovered in any document identified as belonging to a particular language. With this algorithm we propose a method for handling out-of-vocabulary words (OOVs) such that once encountered, OOVs lose their status in future identification runs. In this way, the Spelling Checkers will become smart (intelligent) enough to learn new words. This procedure will involve some human intervention in getting native speakers of the language to confirm words before such words are finally added to the Spellchecker model.

It is clear that the larger the size of the Spelling Checker's lexicon the higher its chances of finding more words, hence the greater the effectiveness of the Spelling Checker. Thus when a document is identified by the Spelling Checker as belonging to a particular language we perform vocabulary extension as follows:

Step 1: Select all unknown words (i.e. OOV words) in the document
Step 2: Get confirmation (from a native speaker of the language) that some or all the words are legitimate words in the particular language.
Step 3: Merge all confirmed words into the Spelling Checker's lexicon.
Step 4: End.

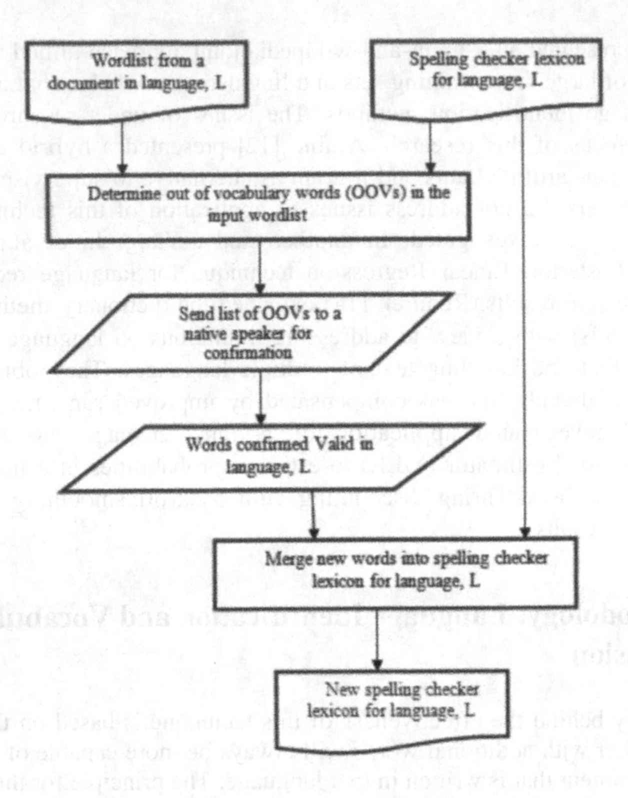

Fig. 1. Language identification with vocabulary extension

The algorithm for vocabulary extension is as follows:

Algorithm for LID with Vocabulary Extension
Input: A newly identified text along with the corresponding Spelling Checker
Output: Spelling Checker with new words added to its lexicon.
1: Load the Spelling Checker and the document (word list)
2: for each word in document
3: if w ∈ Spelling Checker lexicon
4: skip
5: else print w
6: end for
6: # pause and send list (OOVs) to a qualified speaker of the language for confirmation.
7: Return confirmed words for extension of the Spelling Checker's vocabulary.
8: for each confirmed word, w in document
9: wordlist = wordlist + w # (append new word to the vocabulary)
10: end for
11: END.

4 Experimental Setup

In this research, we conducted two experiments. For experiment 1, we used data set in 15 languages comprising nine African languages(Tiv, Hausa, Ndebele, Zulu, Swahili, Igbo, Yoruba, Asante, Akuapem), two Asian languages(Malay, Indonesian), and four European languages(Slovak, Serbian, Croatian, English) from thc UDHR corpus to show that the lexical recall of thcir respective Spelling Checker grows with increasing size of the lexicon. We divided the UDHR corpus translations for the 15 languages into 2 parts, one part consisting of 90% of the corpus while the other part was 10% of the corpus.

We used the 90% part (as training set) to create the Spelling Checker lexicon for each language and then split it into 5 layers (using word frequency) whereby the first layer consisted of words that occur 10 or more times in the training set, the second part consisted of words that occur between 5 and 9 times, the third part was formed by words that occur 3 or 4 times, the fourth part consisted of words that occur twice and the fifth layer consisted of words that occur only once. We tested the Spelling Checker created using the 10% of the corpus by checking first the number of words that is recognizable by the first layer of the Spelling Checker, then the number of words that is recognizable by the first layer plus the second layer and cumulatively added the other layers, testing each time we add one layer to measure how the performance of the Spelling Checker is improving with the increasing number of words in its lexicon. The detail results of our experiments are presented for two of the 15 languages in sections 4.1 and 4.2.

4.1 Results from the Hausa Language Spelling Checker

The Spelling Checker for Hausa language was derived from 90% of the UDHR translation in the Hausa language which had 2,019 words. Upon processing we obtained 443 types of words with the following frequency distribution: 37 words had a frequency of '10 or more', making layer 1 in our categorization, 55 words had frequency of between 5 and 9, 53 words had frequency of 3 or 4, 74 words appeared twice while 224 words appeared only once.

At the first level, i.e. considering words that have frequency of '10 or more' we obtained a low token recall of 49.33% (i.e. percentage of tokens in the testing set that were recognized by layer 1 of the Spelling Checker), and a type recall of 23.81.63%, but the user's recall was as high as 57.33% because to the user, repeated use of words is permitted and significant.

At the second level, 55 words (appearing 5 to 9 times) were added, and the token recall increased to 62.22% while the type recall shifted to 42.86% and the user's recall climbed to 68.00%. The results continued to improve until at the level of adding the hapaxes we found that the token recall climaxed at 87.11%, while the type recall increased to 79.37% and the user's recall finally stood at 88.44%. The result is shown graphically in figure 2 and tabulated in Table 1.

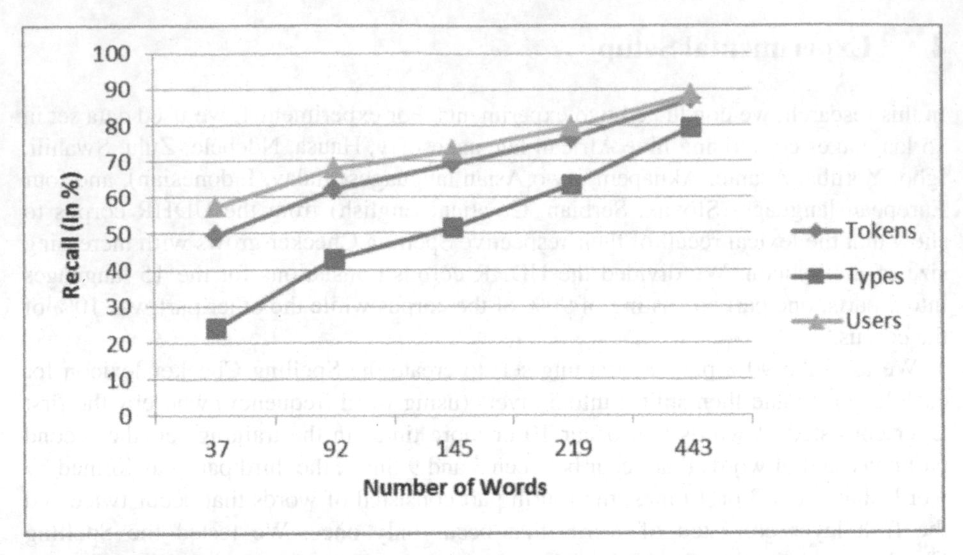

Fig. 2. Graph of recall values for Hausa language Spelling Checker

Table 1. Results of evaluation of the Hausa language Spelling Checker

Spell Checker for Hausa (derived from 90% of UDHR) (2,019 tokens)			10% of UDHR in Hausa (225 tokens, 126 types)				
			Not recognized		**Recall (%)**		
Frequency	**No of words**	**%**	**Tokens**	**Types**	**Tokens**	**Types**	**User's PV**
10 or more	37	8.35	114	96	49.33	23.81	57.33
5 to 9	55	12.42	85	72	62.22	42.86	68.00
3&4	53	11.96	69	61	69.33	51.59	72.89
2	74	16.70	53	46	76.44	63.49	79.56
1(hapaxes)	224	50.56	29	26	87.11	79.37	88.44
Total	**443**	**100**					

4.2 Results from the Tiv Language Spelling Checker

The Spelling Checker for Tiv language was also built using 90% of the UDHR translation in the language with 2,405 words. Here, we got 366 types of words with the following frequency spread: 42 words had a frequency of '10 and above', making layer 1 in our categorization, 47 words had frequency of between '5 and 9', 56 words had frequency of 3 or 4, 56 words appeared twice while 165 words appeared only once.

By testing the first layer we found the token recall was 49.33% while the type recall was 23.81% and the user's recall stood at 57.33%. When 47 words (appearing 5 to 9 times) were included, the token recall increased to 75.00% while the type recall rose to 46.94% and the user's recall climbed to 81.16%. The recall continued to improve until at the level of adding the hapaxes we found that the token recall stood at 89.49% while the type recall rose to 74.49%, and the user's recall finally stood at 90.94%.

4.3 Experimenting on Vocabulary Extension

In order to test the vocabulary extension algorithm, we used the same data set for the 15 languages studied, and split the data set into 3 unequal parts which were used as follows: 70% was used for training, 10% for testing, and 20% for vocabulary extension. The 20% for vocabulary extension was applied by adding 5%, 10%, 15% and 20% to the training set cumulatively while measuring the lexical recall after each step. The results of the experiments are shown in Tables 2, 3 and Figures 3, 4.

Table 2. Results for vocabulary extension for Hausa language

Spell checker for Hausa (Derived from 90% of UDHR - 2,019 tokens)			10% of UDHR for Hausa language - 255 tokens	
Stages of Spell checker lexicon	Types	Tokens	Tokens recognized	% recall
VE Stage1	298	1571	206	80.7%
VE Stage2	351	1683	224	88.0%
VE Stage3	376	1795	232	91.1%
VE Stage4	401	1907	235	92.0%
VE Stage5	443	2019	245	96.0%

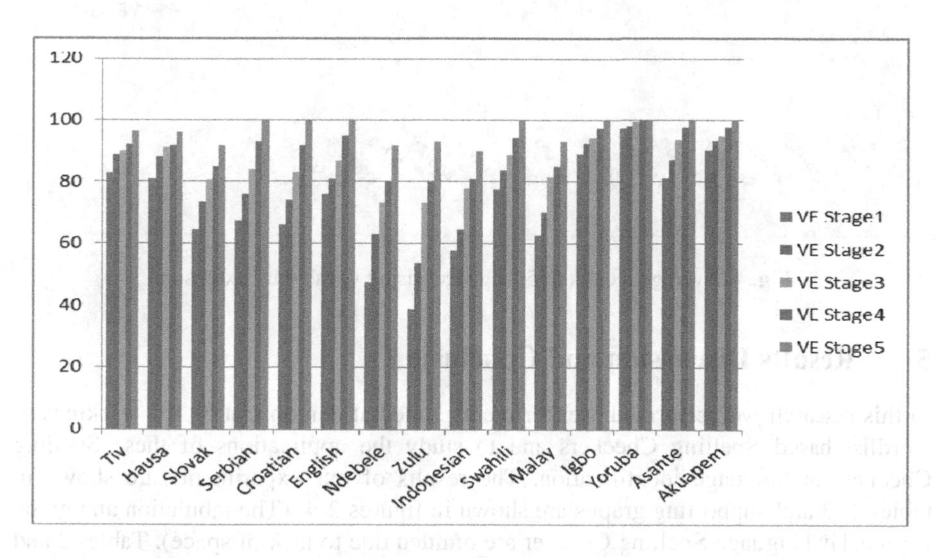

Fig. 3. Average recall of 15 languages on vocabulary extension

Table 3. Results for vocabulary extension for Tiv language

Spell checker for Tiv (Derived from 90% of UDHR – 2,405 tokens)			10% of UDHR in Tiv (276 tokens)	
Stages of Spell checker lexicon	Types	Tokens	Tokens recognized	% recall
VE Stage1	325	1877	228	82.6%
VE Stage2	336	2011	245	88.7%
VE Stage3	340	2145	248	90.0%
VE Stage4	351	2278	257	93.0%
VE Stage5	366	2405	266	96.3%

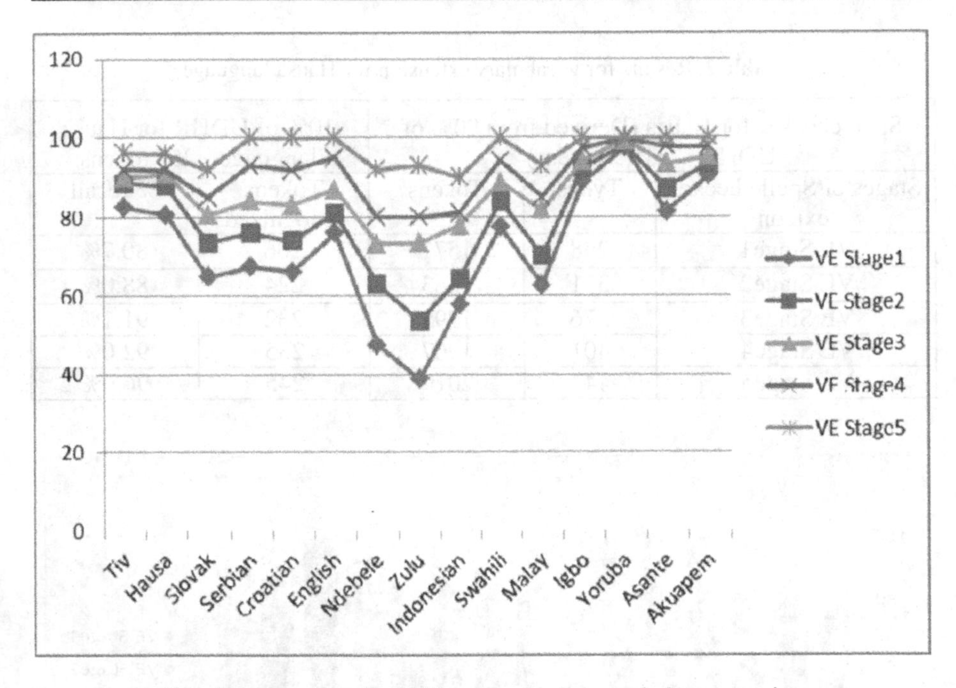

Fig. 4. Average recall of 15 languages using vocabulary extension

5 Results Discussion and Conclusion

In this research we performed 4 experiments aimed at demonstrating the workings of wordlist based Spelling Checkers and to study the applications of these Spelling Checkers in language identification. The results of our experiments are shown in tables 1- 3 and supporting graphs are shown in figures 2-4. (The tabulation and graph for the Tiv language Spelling Checker are omitted due to lack of space). Tables 2 and 3 show the effect of vocabulary extension on recall. 'VE Stage1' means vocabulary extension stage1 and the corresponding columns show the entries accordingly. VE stage1 is the starting point while VE stage2 is obtained by adding 5% of the dataset to the training set and VE stage3 is obtained by adding 10% of the data set. We got VE stage4 and VE stage5 by adding 15% and 20% respectively as explained in section 4.3. 'VE Stage' has the same meaning also in figures 3 and 4.

We have seen that results from all the studied languages confirm that Spelling Checkers have the capability to improve with increasing number of words in their lexica. This suggests that language identification using Spelling Checkers will improve with increasing size of the Spelling Checker lexica. Moreover, the VE algorithm which has no known failure rate is very efficient at extending the vocabulary of Spelling Checkers. Hence vocabulary extension is likely to help in improving the performance of Spelling Checker based language identification. These results suggest that using the Spelling Checker based language identification technique may contribute significantly to the development of digital resources for under-resourced languages. To further improve the performance of the Spelling Checker based language identification we shall consider incorporating optimized search algorithms in future research aimed at enhancing the speed performance of this approach.

Acknowledgments. The authors thank Ministry of Higher Education Malaysia (MOHE) under Fundamental Research Grant Scheme (FRGS) Vot 4F031 and Universiti Teknologi Malaysia under Research University Funding Scheme (Q.J130000.7110.02H47) for supporting the related research.

References

1. Isabelle, P., Foster, G., Plamondon, P.: SILC: un systeme d'identification de la langue et du cadage (1977), http://www-rali.iro.umontreal.ca/ProjectSILC.en.html
2. TextCat Language Guesser, http://www.let.rug.nl/~vannoord/TextCat/ (accessed May 2, 2011)
3. Gordon, R.G.: Ethnologue: Languages of the World, 15th edn. SIL International, Dallas (2005) ISBN: 155671159X
4. Piernaar, W., Snymann, D.: Spelling Checker-based Language Identification for the Eleven Official South African Languages. In: Proceedings of the 21st Annual Symposium of Pattern Recognition of SA, Stellenbosch, South Africa, pp. 213–216 (2011), http://www.prasa.org/proceedings/2010/prasa2010-36.pdf (accessed on December 5, 2011)
5. Prinsloo, D.J., de Schryver, G.M.: Towards Second-Generation Spellcheckers for the South African Languages. In: TAMA 2003 South African Conference Proceedings (2003)
6. Selamat, A.: Improved N-grams Approach for Web Page Language Identification. In: Nguyen, N.T. (ed.) Transactions on CCI V. LNCS, vol. 6910, pp. 1–26. Springer, Heidelberg (2011)
7. Chew, C.Y., Mikami, Y., Nagano, R.L.: Language Identification of Web Pages Based on Improved N-gram Algorithm. International Journal of Computer Science Issues 8(3 (1)) (2011)
8. Office of the High Commissioner for Human Rights 1948- Universal Declaration of Human Rights, http://193.194.138.190/udhr/index.htm (May 2, 2004)

9. Brown, R.D.: Finding and identifying text in 900+ languages. Digital Investigation 9, 34–43 (2012)
10. Vatanen, J., Vayrynen, J., Virpioja, S.: Language Identification of Short Text Segments with N-gram Models (2010), http://www.lrec-conf.org/proceedings/lrec2010/pdf/279_Paper.pdf (accessed on August 2, 2011)
11. Yang, X., Liang, W.: An N-Gram-and-Wikipedia Joint Approach to Natural Language Identification. In: 4th International Universal Communication Symposium (IUCS), Beijing, China, pp. 978–971 (2010), 978-1-4244-7820-0/10
12. Amine, A.B., Elberrichi, Z., Simonet, M.: Automatic Language Identification: An Alternative Unsupervised Approach using a new Hybrid Algorithm. International Journal of Computer Science and Applications 7(1), 94–107 (2010)
13. Yang, J., et al.: Maximum A Posteriori Linear Regression for language recognition. Expert Systems with Applications 39(4), 4287–4291 (2012)
14. Řehůřek, R., Kolkus, M.: Language Identification on the Web: Extending the Dictionary Method. In: Gelbukh, A. (ed.) CICLing 2009. LNCS, vol. 5449, pp. 357–368. Springer, Heidelberg (2009)
15. Ramisch, C.: N-gram Models for Language Detection (2008), http://blog.mohamadzadeh.info/media/blogs/snf/Resource (accessed on August 12, 2011)

Single-Pass Corpus to Corpus Comparison by Sentence Hashing

Dariusz Ceglarek

Poznan School of Banking, Poland
dariusz.ceglarek@wsb.poznan.pl

Abstract. This paper describes a new algorithm identifying common phrase sequences. The $SHAPD2$ algorithm was designed to achieve the goal of a single-pass corpus to corpus comparison. It is a highly efficient solution that finds application with considerable amount of data and excels over other approaches. One of its possible applications is the detection of potential plagiarisms by comparing not a document against a corpus, but corpus to corpus. This makes the $SHAPD2$ algorithm a valuable alternative to the available solutions.

Keywords: plagiarism detection, longest common subsequence, sentence hashing, semantic compression, Natural Language Processing.

1 Introduction

This article presents results of the research aimed at designing a novel algorithm capable of robust detection of common subsequences which is more efficient than the existing methods based on hashing fragments of text (n-grams). The task is to compare 2 large corpora of text documents in a sequential matter – a corpus of suspicious documents against a corpus of source documents (originals) – in order to detect possible plagiarism attempts. The $SHAPD2$ builds upon earlier research activities which produced good results. These solutions were a part of the research on the architecture and functionality of the Semantically Enhanced Intellectual Property Protection System (SeiPro2S) [5]. The novel algorithm proves to be effective in the task defined above, achieving a new level of applicability in previously unreachable scenarios of finding plagiarism cases in big document's corpora almost on-line. Such performance was verified throughout experiments considering millions of document-to-document comparisons.

The SHAPD2 algorithm operates on hash-sums that represent individual sentences. A similar method was used in a number of other known algorithms operating with n-grams, such as $w-shingling$, *minhash*, *simhash*. The first version of $SHAPD$ algorithm was able to compare one suspicious document with document corpus of original documents [6]. However, thanks to improvements, code optimization and introducing new approach, $SHAPD2$ is a new solution, described in detail in the following sections.

A. Bădică et al. (eds.), *Recent Developments in Computational Collective Intelligence*, Studies in Computational Intelligence 513,
DOI: 10.1007/978-3-319-01787-7_16, © Springer International Publishing Switzerland 2014

2 Related Work

The *SHAPD*2 algorithm allows for a robust and a resilient computation of the longest common subsequence shared by one or many input documents. The *SHAPD*2 processes documents by dividing them into a stream of sentences, where unnaturally long sentences (enumerations, itemizations, etc.) are handled by a special procedure [6]. Such an approach allows to extract extremely long sentences from paragraphs and process them individually.

The process is driven by a modular additive hashing function with collision lists. Every concept in a sentence is hashed by assigning a number from a previously defined range. Furthermore, the individual hashes are summed to represent a sentence. Thanks to the additive nature of the hashing function, sentences with a changed concept order are treated as equivalents. Thus, the resulting algorithm not only finds the longest common sequences, but also the longest common quasi-sequences (allowing minor editing changes such as syntactic changes, insertions, deletions and synonym replacements, as well as combining or splitting multiple sentences to change their structure).

The task of matching the longest common subsequence is an important one in many subdomains of computer science. Its most naive implementation was deemed to have a time complexity of $O(m_1 * m_2)$ (where m_1 and m_2 are the numbers of concepts in compared documents). One of the most important implementations of the search for the longest common subsequence can be found in [13], which features time complexity $O((m_1 * m_2)/log(m_2))$. This is the fastest algorithm that does not operate with text frames and their hashes. All of the above cited works use algorithms whose time complexity is near quadratic which results in drastic drop of efficiency when dealing with documents of considerable length.

It was first observed in [15] that the introduction of a special structure based on a hashing technique which was later known as shingling or chunks (a continuous sequence of fixed-length tokens in a document) can substantially improve the efficiency determining the similarity level of two documents by observing the number of common shinglings. This technique was introduced to detect near duplicate web pages. The following works such as [3,2] introduce further extensions to the original idea. Charikar [8] proposed a locality sensitive hashing scheme for comparing documents. A number of works represented by publications such as [3] or [1] have provided plausible methods to further boost the measuring of similarities between entities. Later, Henzinger [11] combined the algorithms of Broder et al. and Charikar to improve overall precision and recall.

The important distinction between solutions described above and the SHAPD (version 1 and version 2) is the emphasis on a sentence as the basic structure for comparison of documents and a starting point of determining a longest common subsequence. Thanks to such an assumption, SHAPD (version 1 and version 2) provides better results in terms of time needed to compute the effects. Moreover, its functioning does not end at the stage of establishing that two or more documents overlap. It readily delivers data on which sequences overlap, the length of the overlapping and it does so even when the sequences are locally discontinued.

The capability to perform these makes it a method that can be naturally chosen in plagiarism detection, because such situations are common during attempts to hide plagiarism. In addition, it implements the construction of hashes representing the sentence in an additive manner, thus word order is not an issue while comparing documents.

The $w - shingling$ algorithm runs significantly slower when the task is to give length of all long common subsequences. Due to the fixed frame orientation, when performing such operating $w - shingling$ behaves in a fashion similar to the Smith-Waterman algorithm resulting in a significant drop of efficiency.

The importance of plagiarism detection is recognized in many publications. It may be argued that, it is an essential task in times, when access to information is nearly unrestricted and a culture for sharing without attribution is a recognized problem (see [16] and [4]).

3 The $SHAPD2$ Algorithm

Hashing is a technique commonly used in NLP tasks used in order to achieve faster word retrieval. In plagiarism detection it is crucial, however, to identify and match longer common word sequences (with special focus on sentences).

Nevertheless, there are situations in text documents where long passages of text stretch for dozens of lines before reaching a full-stop mark (such as different types of enumerations, tables, listings, etc.). Some sort of strategy needs to be devised for such cases, i.e. how to split portions of text, which are longer than the reasonable length of a sentence in a natural language.

$SHAPD2$ utilizes a brand new mechanism to organize the hash-index as well as to search through the index. It uses additional data structures such as correspondence arrays to aid in the process.

As introduced before, there are two corpora of documents which comprise the algorithm's input: a corpus of source documents (originals) $D = \{d_1, d_2, ..., d_n\}$, and a corpus of suspicious documents to be verified regarding possible plagiaries, $P = \{p_1, p_2, ..., p_r\}$. $SHAPD2$ focuses on whole sentence sequences. A natural way of splitting a text document is to divide it into sentences and it can be assumed that documents containing the same sequences also contain the same sentences.

Before applying algorithm for there is necessary to carry out text-refinement process what is standard procedure in NLP/IR task (starting from unstructured text document input to a structure containing stacked sequentially descriptors of concepts found in the input document). Action that make up the process of text-refinement in documents starts from extracting lexical units (tokenization), and further text-refinement operations are: elimination of the words without semantic importance from the so-called information stop-list, the identification of multiword concepts (when phrase of several words create one concept), bringing concepts to the main form by lemmatization (for Polish documents) or stemming (for English documents using a popular Porter stemmer [18]). It is particularly difficult task for flexible languages, such as Polish or French (multiple noun

Table 1. Sample hash key determination in *SHAPD*2

dayan-92.pdf
The Convergence of TD(X) for General X PETER DAYAN *Machine Learning, 8, 341-362 (1992), 1992 Kluwer Academic Publishers, Boston*
The methods of temporal differences (TD), first defined as such by Sutton (1984; 1988), fall into this simpler category. Given some parametric way of predicting the expected values of states, they alter the parameters to reduce the inconsistency between the estimate from one state and the estimates from the next state or states.

hash value	sentence/frame
2906878	methods temporal differences td defined sutton fall simpler category
2496872	given parametric way predicting expected values states alter
2339613	parameters reduce inconsistency estimate state estimates state states

declination forms and verb conjugation forms). In lemmatization procedure there is used Ispell dictionary for Polish documents and finite state automaton. The goal of both stemming and lemmatization is to reduce inflectional forms and sometimes derivationally related forms of a word to a common base form.

Synonyms need to be represented with the same concept descriptors using lexical relationships of synonymy from semantic network. It allows correct similarity analysis and also increases classification algorithms efficiency without loss in comparison quality [12].

Abstracting process faces another problem here, which is polysemy. One word can represent multiple meanings, so the apparent similarity need to be eliminated. It is done by concept disambiguation (using local analysis), which identifies word meaning depending on its context, is important to ensure that no irrelevant documents will be returned in response to a query [14].

The last operation in the text-refinement procedure of the SeiPro2S system may be the generalization of concepts by using semantic compression). The final effect of refinement procedure is the structure of documents containing ordered descriptors of concepts derived from the input document. This structure can be stored as an abstract (data for creating index) of the document, and then use during phase 2 (comparing documents). Sample outputs from a sentence splitting and hash calculation in Phase 1 are shown in Table 1.

Then in Phase 1 oa SHAPD2 algorithm, all documents need to be split into text frames of comparable length – preferably sentences, or in the case of longer sentences – shorter phrases. A coefficient α is a user-defined value, which allows to set the expected number of frames that a longer sentence is split into. The new procedure of uniform fragmentation is described in listing of Algorithm 1. As a result, every document from the original corpus, as well as all suspicious documents, is represented by index as a list of sentence hashes calculated separately for all concepts and summed for each frame.

As an output of the text-refinement process the system produces vectors containing ordered concept descriptors coming from sentences from the input document. The first version of the SHAPD algorithm [6] was able to compare a

Algorithm 1. Phase 1. Splitting text form document d_i into comparable frames and calculate hash key values for frames

$f := round_u p(l/\alpha)$
while $f > 0$ **do**
  $a := round_u p(l/f)$
  $\bar{c_f}$:= getConceptsFromSentence(s, a)
  $k_{i,f}$:= calculateHash($\bar{c_f}$)
  $f := f - 1$
end while
l - sentence s length
α - alpha coefficient
f – number of frames to split a longer sentence into
a – current frame length
$\bar{c_f}$ – vector of concepts from f-th frame
$k_{i,f}$ – hash key for frame f from document d_i

suspicious document with exactly one document from the corpus P. The SHAPD2 algorithm is able to compare a suspicious document with the entire corpus of source documents in a single pass.

For all documents p_i from corpus P (containing suspicious documents), the correspondence array CL and maxima array TM are cleared. For each frame, set of tuples is retrieved from index table T. If there are any entries existing, it is then checked whether they point to the same source document and to the previous frame. If the condition is true, the local correspondence maximum is increased by one. Otherwise, the local maximum is decreased.

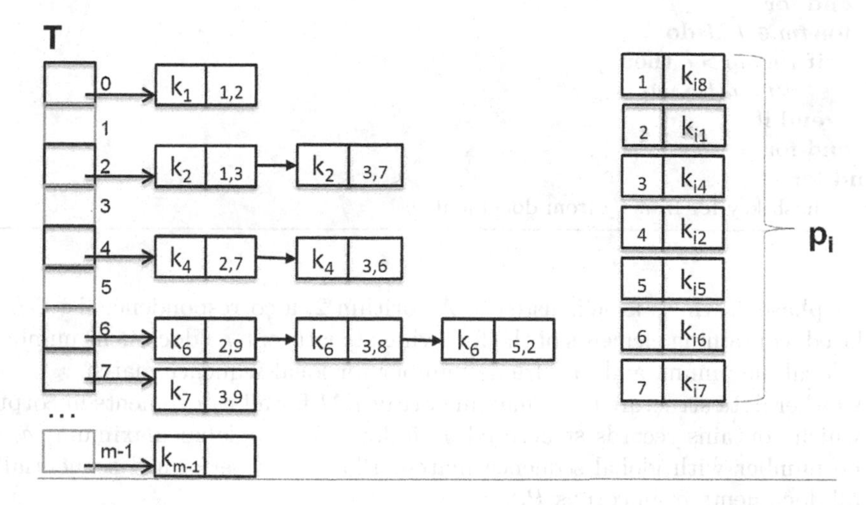

Fig. 1. Hash table T indexing a corpus D of source documents and hashes from suspicious document $d_i \in P$. Details are presented in Table 2.

After all of the frames are checked, table TM storing the correspondence maxima is searched for records whose correspondence maxima are greater than a threshold set e (the number of matching frames to be reported as a potential plagiarism). Frame and document number are returned in these cases.

In the next step, a hash table T is created for all documents from corpus D, where for each key the following tuple of values is stored: $T(k_{i,j}) =< i, j >$, (document number, frame number) (see Figure 1).

Algorithm 2. Phase 2

```
{Build hash table T}
for d_i ∈ D do
   for frame_j ∈ d_i do
      addHashToIndex(T)
   end for
end for
{Find longest common frame sequences}
for p_i ∈ P do
   clear(CL)
   clear(TM)
   for frame_j ∈ p_i do
      if exists T(k_{i,j}) then
         if T(k_{i,j}).i = i ∧ T(k_{i,j}).j = j-1 then
            CL(i).ml ++
            update(TM)
         end if
      else
         CL(i).ml --
      end if
   end for
   for tm ∈ TM do
      if tm.mg > e then
         return(tm.ng)
      end if
   end for
end for
k_ij - hash key for frame j from document p_i
```

For phase 2, whose logic is listed in Algorithm 2, a correspondence list CL is declared, containing elements of the following structure: n_d – document number, m_l – local maximum, and n_l – frame number for local sequence match.

Another data structure is the maxima array TM for all r documents in corpus P, which contains records structured as follows: m_g – global maximum, n_g – frame number with global sequence match. Phase 2 is performed sequentially for all documents from corpus P.

Table 2. Details of comparison (presented in Figure 1) between a suspicious document p_i and a corpus D represented by hash table T

$k_i \in T$	CL table	TM table
1 $k_8 \notin T$		
2 $k_1 \in T$	$\{m_l = 1, n_d = 1, n_l = 2\}$	$TM_1 = \{m_g = 1, n_g = 2\}$
3 $k_4 \in T$	$\{m_l = 1, n_d = 2, n_l = 7\}$	$TM_2 = \{m_g = 1, n_g = 7\}$
	$\{m_l = 1, n_d = 3, n_l = 6\}$	$TM_3 = \{m_g = 1, n_g = 6\}$
4 $k_2 \in T$	$\{m_l = 1, n_d = 1, n_l = 3\}$	
	$\{m_l = 2, n_d = 3, n_l = 7\}$	$TM_3 = \{m_g = 2, n_g = 7\}$
5 $k_5 \notin T$	$\{m_l = 1, n_d = 3, n_l = 7\}$	
6 $k_6 \in T$	$\{m_l = 1, n_d = 2, n_l = 9\}$	
	$\{m_l = 2, n_d = 3, n_l = 8\}$	
	$\{m_l = 1, n_d = 5, n_l = 2\}$	$TM_5 = \{m_g = 1, n_g = 2\}$
7 $k_7 \in T$	$\{m_l = 3, n_d = 3, n_l = 9\}$	$TM_3 = \{m_g = 3, n_g = 9\}$

k_j – hash key of frame j from document d_i

4 Experiment

In order to evaluate the algorithm's efficiency, a series of experiments were carried out. The most widely used test collection of annotated documents Reuters-21578 [19] for text categorization has been used as a collection of source documents as well as suspicious documents (including potential plagiarism cases) derived from the Reuters-21578 corpus.

A set of 3000 original documents was used as source set, and several sets including 1000 to 6000 suspicious documents were used as a set of suspicious documents. The sets were compared using two algorithms: $w - shingling$ and $SHAPD2$. w-shingling algorithm represents a family of n-grams methods containing also minhash, simhash algorithms. All tests were carried out on one computing platform, a stock laptop computer with an 8-core processor, clocked at 2.0 GHz.

The basic results of efficiency test are as follows. A comparison of one suspicious document to a set of 3000 originals (containing about 3060 words on average) takes 7.13 ms. As many as 420,700 document-to-document comparisons were achieved in 1-second intervals.

The resulting times of tests' execution (presented in Table 3) include the loading time of the prevoiusly created indexes and the time of comparing suspicious documents with whole corpus P. It can be observed that the efficiency turns

Table 3. Processing time [ms] for comparing n suspicious documents with a corpus of 3,000 original documents

n	1000	1500	2000	2500	3000	3500	4000	4500	5000	5500	6000
w-shingling	5680	6654	8581	9478	11967	14864	16899	20242	23200	33955	50586
SHAPD2	4608	5114	5820	6374	7125	7213	7527	7818	8437	8656	8742

Table 4. *SHAPD*2 and *w – shingling* methods' performance on the PAN-PC-10 plagiarism corpus. The metrics used - precision, recall, granularity and overall mark: *plagdet* - correspond to the International Competition on Plagiarism Detection scores).

Method	PlagDet	Precision	Recall	Granularity
SHAPD2 (original text)	0.626	0.979	0.401	1.00
SHAPD2 (semantic compression 1500)	0.691	0.505	0.945	1.00
w-shingling (4-grams)	0.615	0.949	0.399	1.00
w-shingling (6-grams)	0.601	0.962	0.381	1.00

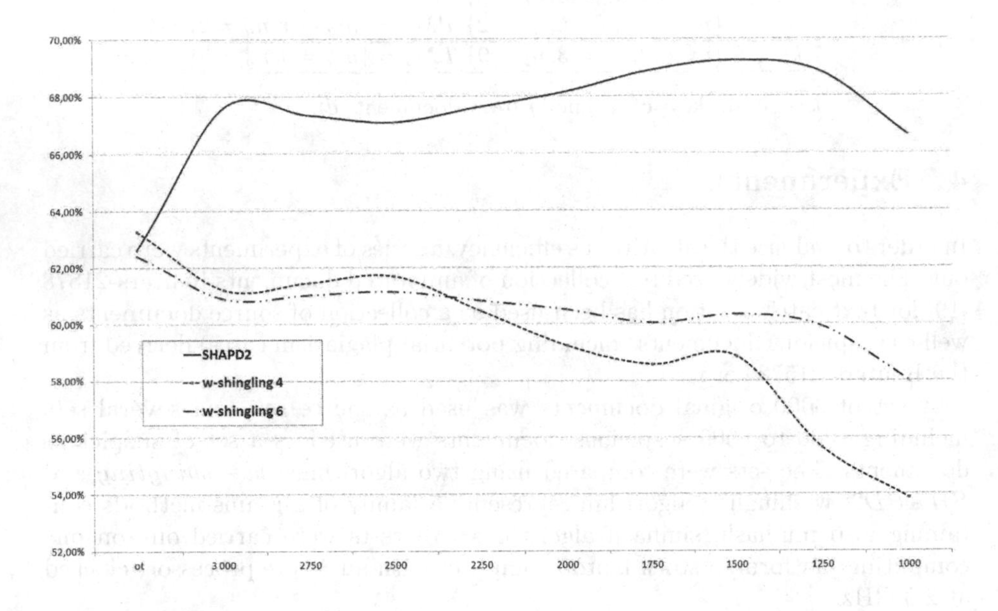

Fig. 2. Comparison of the synthetic plagdet indicator: *SHAPD*2 versus *w – shingling* 4-grams and 6-grams - for all cases of plagiarism in the PAN-PC corpus

out to be better by an order of magnitude from *w – shingling*. This enables plagiarism detection using much larger source document corpora.

Moreover, the algorithm uses the same techniques whose high effectiveness has already been proven in plagiarism detection employing semantic compresion, as well as their strong resilience to false-positive examples of plagiarism[7], which may be an issue in cases where competitive algorithms are used. The Clough & Stevenson Corpus of Plagiarised Short Answers[9] was used for the benchmark.

The PAN-PC-10 plagiarism evaluation corpus[1]from Weimar University has been utilised in order to run a benchmark. PAN corpora have been used since

[1] `http://pan.webis.de/`

2009 in plagiarism uncovering contests, gathering researchers from this domain to evaluate their methods in comparison with others. The possibility to use the same data sets and measures allows to perceive the results as reliable (for details see Table 3). Figure 4 shows that the SHAPD2 algorithm produces both better results of recall and a synthetic plagdet indicator than $w - shingling$. However, the results of the first attempt of using SHAPD2 to detect plagiarism in the PAN-PC-10 corpus are still below the best system in the original PAN-PC-10 task (achieved results of .797 of *plagdet* indicator).

5 Conclusions

To summarize, the following should be emphasized:

- The algorithm developed has a very low computational complexity – evaluated in experiments to linearithmic, proving to be extremely efficient.
- $SHAPD2$ algorithm is resilient to fluctuating word order in sentences. It is especially important in cases of languages with a highly flexible syntax (eg. Polish, which allows multiple correct word sequences, although having a Subject-Verb-Object-based syntax, like English).
- The algorithm is resilient to small sentence inclusions or deletions (as the S&W algorithm is), which is an important feature in plagiarism detection, as it is a common strategy of slight modifications when committing plagiaries.
- Utilization of NLP techniques, such as concept identification and disambiguation, semantic compression, surely improve the effectiveness of plagiarism detection, which is subject to further developments.
- The designed $SHAPD2$ algorithm - employing semantic compression - is strongly resilient to false-positive examples of plagiarism which may be an issue in cases when competitive algorithms are used.

In the near future, author plans to further develop various algorithms and reorganize of available assets so that the semantic compression can be applied in a automated manner to text passages without introduction of hypernyms disrupting user's experience. Next step is applying FrameNet approach [10] for establishing equivalent phrases not detecting by semantic compression mechanism.

References

1. Hamid, O.A., Behzadi, B., Christoph, S., Henzinger, M.: Detecting the origin of text segments efficiently. In: Proceedings of the 18th International Conference on World Wide Web, WWW 2009, vol. 7(3), pp. 61–70 (2009)
2. Andoni, A., Indyk, P.: Near-optimal hashing algorithms for approximate nearest neighbor in high dimensions. Commun. ACM 51(1), 117–122 (2008)
3. Broder, A.Z., Glassman, S.C., Manasse, M.S., Zweig, G.: Syntactic clustering of the web. Comput. Netw. ISDN Syst. 29(8-13), 1157–1166 (1997)
4. Burrows, S., Tahaghoghi, S.M.M., Zobel, J.: Efficient plagiarism detection for large code repositories. Softw. Pract. Exper. 37, 151–175 (2007)

5. Ceglarek, D.: Architecture of the Semantically Enhanced Intellectual Property Protection System. In: Burduk, R., Jackowski, K., Kurzynski, M., Wozniak, M., Zolnierek, A. (eds.) CORES 2013. AISC, vol. 226, pp. 711–720. Springer, Heidelberg (2013)
6. Ceglarek, D., Haniewicz, K.: Fast plagiarism detection by sentence hashing. In: Rutkowski, L., Korytkowski, M., Scherer, R., Tadeusiewicz, R., Zadeh, L.A., Zurada, J.M. (eds.) ICAISC 2012, Part II. LNCS, vol. 7268, pp. 30–37. Springer, Heidelberg (2012)
7. Ceglarek, D., Haniewicz, K., Rutkowski, W.: Robust Plagiary Detection Using Semantic Compression Augmented SHAPD. In: Nguyen, N.-T., Hoang, K., Jędrzejowicz, P. (eds.) ICCCI 2012, Part I. LNCS, vol. 7653, pp. 308–317. Springer, Heidelberg (2012)
8. Charikar, M.S.: Similarity estimation techniques from rounding algorithms. In: Proceedings of the Thiry-fourth Annual ACM Symposium on Theory of Computing, STOC 2002, pp. 380–388. ACM (2002)
9. Clough, P., Stevenson, M.: A Corpus of Plagiarised Short Answers (2009), http://ir.shef.ac.uk/cloughie/resources/plagiarism_corpus.html (access: April 2, 2012)
10. Goddard, C., Schalley, A.C.: Semantic Analysis. In: Indurkhya, N., Damerau, F. (eds.) Handbook of Natural Language Processing, pp. 93–121. Chapman & Hall/CRC (2010)
11. Henzinger, M.: Finding near-duplicate web pages: a large-scale evaluation of algorithms. In: SIGIR 2006: Proceedings of the 29th Annual International ACM SIGIR Conference on Research and Development in Information Retrieval, pp. 284–291. ACM, New York (2006)
12. Hotho, A., Staab, S., Stumme, G.: Explaining Text Clustering Results Using Semantic Structures. In: Lavrač, N., Gamberger, D., Todorovski, L., Blockeel, H. (eds.) PKDD 2003. LNCS (LNAI), vol. 2838, pp. 217–228. Springer, Heidelberg (2003)
13. Irving, R.W.: Plagiarism and collusion detection using the Smith-Waterman algorithm, Technical report, University of Glasgow, Glasgow (2004)
14. Krovetz, R., Croft, W.B.: Lexical Ambiguity and Information Retrieval (1992)
15. Manber, U.: Finding similar files in a large file system. In: Proceedings of the USENIX Winter 1994 Technical Conference, WTEC 1994, p. 2. USENIX Association, Berkeley (1994)
16. Ota, T., Masuyam, S.: Automatic plagiarism detection among term papers. In: Proceedings of the 3rd International Universal Communication Symposium, IUCS 2009, pp. 395–399. ACM, New York (2009)
17. PAN-PC-10 Corpus on line, http://pan.webis.de/ (access March 11, 2013)
18. Porter, M.F.: An algorithm for suffix stripping. Program 14(3), 130–137 (1980)
19. Reuters-21578 Corpus, Carnegie Group, Inc. and Reuters, Ltd. (2004), http://www.daviddlewis.com/resources/testcollections/reuters21578 (access April 10, 2012)

Opinion Classification in Conversational Content Using N-grams

Kristina Machova and Lukáš Marhefka

Dept. of Cybernetics and Artificial Intelligence, Technical University, Letná 9, 042 00, Košice, Slovakia
kristina.machova@tuke.sk

Abstract. The paper introduces the problem of opinion classification related to conversational content. It describes briefly various approaches known in this field. The focus is on a novelty method which has been designed on the basis of cyclic usage of n-grams (4-grams). This method belongs to lexicon based approaches. The contribution describes implementation of this method for the Slovak language, test results of the presented implementation and discussion of the achieved results as well.

Keywords: Opinion mining, conversational content, web discussions, n-grams.

1 Introduction

It is well known that the social web increases level of interaction between users mainly regarding to conversational content (chats, discussion forums, blogs and so on). Within these interactions, people (web users) influence each other. This influence concerns decision making about various life situations, for example decisions about purchase or production and selling some products, decisions about voting political representatives and so on.

Our motivation is to design an opinion classification application to support such decision making. This application is expected to mine opinions from conversational content in an automatic way – it should free users from reading all related discussions and sometimes great number of contributions. Such application can be valuable from the following point of view. It mines opinions from web conversational content, which has been produced spontaneously and therefore it contains real, truthful and unaffected opinions of users – opinions both critical as well as complimentary.

The conversational content includes positive or negative opinions, attitudes or views of persons, who play the roles of opinion owners. Opinions are related to a given entity or some of its parts. The entity is some product, person, event, organization, etc. It is the object, we are talking about. Our approach, presented within this paper, is focused on the problem of opinion polarity classification, which represents determination (calculation) of the polarity (positive or negative) of a subjective opinion.

A. Bădică et al. (eds.), *Recent Developments in Computational Collective Intelligence*,
Studies in Computational Intelligence 513,
DOI: 10.1007/978-3-319-01787-7_17, © Springer International Publishing Switzerland 2014

The most similar to our approach is [8]. It uses a dictionary of words annotated with their orientation (polarity). This work splits this dictionary into four sub-dictionaries according to word classes (adjectives, nouns, verbs and adverbs), and these dictionaries are checked for consistency and reliability. According to [8], more words in the dictionary can lead to noise increase and subsequently to precision decrease.

Another approach presented in [9] is focused on the SentiStrength detection algorithm, which solves some problems connected with sentiment analysis (generation of the sentiment strength list, optimization of the sentiment word strengths, allocation of the miss words, spelling correction, creation of the booster word list, the negating word list and emoticon list, repeated letters processing and ignoring negative emotion in questions). Their approach is based on using machine learning techniques.

Paper presented within [1] is focused on sentiment analysis. It presents a novel learning based approach that incorporates inference rules inspired by compositional semantics into the learning procedure. Design of this method represents meaning of a composed expression as a function of the meanings of its parts within the compositional semantics. The approach presented in [1] processes negations and intensification separately. This processing is made over the whole text with the aid of "voting-based inference". Some effort was spent on semantically enriching algorithms for analysis of web discussion contributions by authors of [5]. Also dedicated information can be used as an interface to newsgroup discussions [6].

2 Opinion Classification within Conversational Content

There are different approaches to extracting sentiment (opinion) automatically according to [8]:

- Lexicon based approach, which calculates the polarity of a contribution (document) from polarities of words or phrases drawn from the contribution. According to [4] the sentiment lexicon generation can involve a dictionary-based approach and a corpus-based approach.
- Classification based approach – builds classifiers from labeled instances of contributions or sentences. This approach could be described as a statistical or machine learning approach.

2.1 Classification Based Approach

Within the classification based approach, many of well-known machine learning methods can be used, for example Naïve Bayes classifier or Support Vector Machines (SVM), as well as statistical methods, for example Maximal Entropy and so on. All these algorithms were used in [2] focused on the automatic classification of opinions from the micro-blog service Twitter.

Machine learning methods estimate users' opinion on the basis of an annotated training sample consisting of training examples. These training examples can be obtained by-the creation of annotations of conversational contributions from the social web. An important drawback of machine learning approach is its dependency on huge amount of annotated training documents. In our problem, considerable amount of annotations of training contributions to web discussions is needed. To facilitate the creation of these annotations, an annotation tool with some level of automation can be utilized. The source [7] introduces some existing annotation tools, for example GATE (General Architecture for Text Engineering), SemTag (Semantic annotation Tool), annotation platform KIM (Knowledge and Information Management) and Luvak – more general semi automatic annotation tool built on Eclipse platform.

2.2 Lexicon Based Approach

This approach looks on the input text as a set of words. It does not take into account relationships between words within sentences or grammatical rules. Not every word from processed text has the same importance within the opinion classification. Our focus is on words, which can express opinion in the best way. Such words are stored in the classification dictionary. On the base of known polarity of such characteristic words, polarity of whole texts from conversation and finally polarity of whole web discussion can be determined. The simplest dictionaries are the dictionaries which enable binary classification. More suitable are fuzzy dictionaries, which are able to determine not only polarity but also the strength of the polarity. Such dictionary can be created for only a given application. But also some known lexicon can be used, for example Word Net, WordNet-Affect, SenticNet, SentiWordNet and so on.

Our approach, presented within this contribution is the lexicon based approach. It contains words, which are common within web conversation forums including slang words, words without diacritics and words with more common grammatical mistakes.

Besides basic problems of opinion analysis as word subjectivity identification, word polarity (orientation) determination and determination of intensity of the polarity, each application of opinion classification has to solve the processing of negation and intensification. The basic problems can be simply solved with the aid of classification dictionaries. These dictionaries focus on those words, which are able to express subjectivity very well - mainly adjectives (e.g. 'extraordinary') and adverbs (e.g. 'awfully') are considered. On the other hand, other word classes must be considered as well in order to achieve satisfactory precision, for example nouns (e.g. 'crash') or verbs (e.g. 'damage'). The words with subjectivity are important for opinion classification. They are identified and inserted usually into a set of vocabularies – one vocabulary for each word class (adjectives dictionary, adverbs dictionary, nouns dictionary and verbs dictionary). Words with subjectivity are inserted into the corresponding vocabulary together with their degree of polarity.

2.3 Negation and Intensification Processing

There are two basic approaches to *negation processing*: switch negation and shift negation. The *switch negation* is simply reversion of the polarity of the lexical item. The reversion represents changing the number sign of the polarity degree (from minus to plus and vice versa). There are many various words related to negation that need to be taken into account: *not, none, never, nothing*, which usually are situated next to a related word – item. But also other negations should be taken into account as *without, don't, lack* and so on, which can be situated in significant distance from the lexical item. These negations can be hardly processed by switch negation. What is more, switch negation can be insufficiently precise, because negation of a strong positive word rarely is a strong negative word and vice versa. More often the negation of a strong positive word is a slightly negative word and vice versa.

The *shift negation* focuses on the case, when negation of a strong positive (negative) word is not strong negative (positive). Instead of changing the sign, "shift negation" shifts polarity degree toward the opposite polarity by a fixed value (value 4 in implementation 11]). Thus "a + 2" adjective is negated to "a –2", but the negation of "a – 3" adjective is only slightly positive "a + 1". For example: "She's not terrific (5 – 4 = 1) but not terrible (–5 + 4 = –1) either."

The *intensification processing* supposes the dictionary of intensifications – words, which are able to increase (or decrease) the intensity of polarity. According to [8], intensifiers can be of two categories: *amplifiers* (e.g., *very*) increase the semantic intensity of a neighboring lexical item, whereas *downtowners* (e.g., *slightly*) decrease it. Each of intensifications is stored in dictionary together with sign and number. This number represents a percentage of changing of polarity intensity and sign represents the type of this change (+ represents increasing of polarity value by amplifier and – represents decreasing of polarity value by downtowner).

3 N-grams in Opinion Classification

Before the opinion classification step, an input text preprocessing step must be provided. This text preprocessing represents some transformation of discussion texts into lexical units which can be subsequently easily processed. There are several approaches how to parse an analyzed input text into smaller lexical units. The most used approaches are n-grams and part-of-speech tagging.

N-gram can be defined as a series of items from some sequence. From the semantic point of view, it can be a sequence of phones, characters or words. In practice, n-gram as a sequence of words is the most common. The sequence of two (three) words is called bigram (trigram). For the case, when more than three words (exactly *n*) are in the same sequence, such sequence is called n-gram. N-grams are used in the wide scale of fields, as theoretical mathematics, biology, cartography, even in the field of music. In the field of natural language processing, n-grams can be used for words prediction. The words prediction uses so called "n-grams model". This n-gram model calculates the probability of occurrence of the last word in an n-gram from the previous n-grams. Another way of using n-grams is plagiarism discovery by text dividing into smaller fragments. These fragments are represented by n-grams. These n-grams can be easily compared and consequently, the measure of similarity of comparing documents can be calculated. N-grams are often used for text categorization and also for effective searching for correct candidates of misspelled words. Our approach, presented within this contribution uses n-grams for splitting the web discussion contributions into lexical units.

The part-of-speech (POS) tagging represents the recognition of word class within the text on the basis of the given language attributes and relationships between word classes. The POS tagger is a program, which is able to read text and to assign word class to each word from the given text. Such word class assigning can be based on the word definition or the word position within the sentence. There are three types of POS taggers: rule taggers, stochastic taggers and transformation taggers. The rule tagging algorithms are based on sets of rules, which are used for tagging of the processed text. This tagged text can be used as a training set for stochastic approach. The stochastic taggers are based on probability of the given tag occurrence within the given text.

3.1 Design of N-grams Approach to Opinion Extraction

Our n-grams application to opinion classification belongs to lexicon based approaches. This application is oriented on the Slovak language, so it uses a dictionary of Slovak language words. The structure of this dictionary is given by four more important word classes for opinion analysis: adjectives, adverbs, nouns and verbs. The dictionary consists of two different parts. The first part contains adjectives, nouns and verbs. The second one contains adverbs and negations. The first part of the dictionary is used for solving basic problems of opinion classification. The second part of the dictionary is used for negation processing and intensification because the adverbs have in a language the function to increase (*"surprisingly nice"*) or to decrease (*"extremely low-class"*) intensity of a word polarity. The dictionary contains also some emoticons, which naturally can express emotions and opinions very well. Sometimes, the analyzed text can be less clear and so emoticons can increase the precision of the classification. These emoticons are stored within the first part of the dictionary. They are illustrated in Table 1.

Table 1. Emoticons with positive and negative polarity

Positive	Negative
:)	:(
:))	:((
:)))	:(((
:-)	:-(
=)	=(
:D	
=D	

All words and emoticons from this first dictionary are quantified to polarity degree within the interval from -3 to 3. Some illustrative examples of these quantifications are presented in Table 2. With respect to the second part of the dictionary, intensifiers (adverbs) are assigned by values from -0.5 to 1 and negations are represented by value -2. This is illustrated in Table 3.

Table 2. Polarity degrees of example words and emoticons (the first part of the dictionary)

Polarity degree	Words and emoticons
3	:D, godlike, extra ordinal
2	:), super, excellent
1	nice, functional, OK
-1	unpleasant, weak
-2	:(, shocking, miserable
-3	:((, fatal, catastrophic

Table 3. Polarity degrees of negations and intensifications (second part of the dictionary)

Polarity degree	Words
1	very, totally, extraordinary
0.5	suitably, really, actually
-0.5	little, overly, unnecessarily
-2	Negations: no, not, don't...

The analyzed text is processed by the following way. All words from the text are compared with all words stored in the first and second parts of the dictionary (shortly the first dictionary and the second dictionary). In the case of match within the first dictionary, the values of all matching words from this dictionary are summed. The resulting sum is solution of the basic problems of opinion analysis and represents the first sum in the formula (1). Within this first sum, only polarity values of adjectives, nouns and verbs are considered. The second sum takes into account negations and intensifications of related words, which have been incorporated within the first sum. This is the reason, why we decided to use multiplication (not aggregation) as the operation between the first sum and the second sum. The second sum aggregates the values obtained from the second dictionary (for intensifiers and negations). But in the case of contributions, which do not contain any negation or intensification, the second sum would be zero and consequently the resulting polarity of the contribution would be zero. Thus, value "1" was added to the second sum as a neutral value. This idea is represented by formula (1):

$$P = \sum v(w_i^1)[1+\sum v(w_j^2)] \ . \tag{1}$$

where:

P ...is the polarity degree of analyzed text

$v(w_i^1)$ ...is the value of word w_i of text found in the first part of the dictionary

$v(w_j^2)$ ...is the value of word w_j of text found in the second part of the dictionary

$\sum v(w_i^1)$...is the first sum from the dictionary (solution of the basic problem)

$\sum v(w_j^2)$...is the second sum from the second dictionary (solution of the negation and intensification).

For example:

A sentence containing only positive and negative words without negations and intensifications – it can be processed using only the first part of the dictionary

- the lexical unit *"The mouse is nice but the processing is miserable and globally it is unsuccessful."* is processed by the following way:
  nice (+1) + miserable (-3) + unsuccessful (-1): P = -3

A sentence containing negation

- the lexical unit *"It is not good solution."* is processed by the following way:
  good (+1) first dictionary, not (-2) second dictionary: P = 1*[1+(-2)] = -1

A sentence containing intensification (very)

- the lexical unit *"Globally, the processing is very polite."* is processed by the following way: polite (+1) first dictionary, very (+1) second dictionary: P = 1*[1+1)] = +2.

The contribution of the presented approach in comparison with a similar approach [8] is the following. The application presented in [8] splits its dictionary into four sub-dictionaries according to word classes (adjectives, nouns, verbs and adverbs). On the other hand, we use only two sub-dictionaries, one with adjectives, nouns and verbs to support solving basic problems of opinion classification and the second one with adverbs and negations for processing intensification and negation. In our approach the dictionary can be created directly from web discussions, which increases the precision of opinion analysis. The originality of our approach in comparison with [8] is mainly in usage of a different method for negation and intensification processing based on formula (1). Our approach to intensification and negation processing is briefly characterized in Table 3 and by examples following the formula (1) within the 3.1 section. On the other hand, our tests were not as complex as in [8].

In [8], intensification is provided by increasing (decreasing) the semantic intensity of neighbouring lexical items using a special dictionary of intensifiers annotated with percentage of polarity increasing (decreasing) of related words. Our design, based on the formula (1), doesn't work with percentages. The Table 4 presents a discrete scale. Its first version was {1, 0.5, -0.5, -1}, which represented the measure of increasing (1, 0.5) or decreasing (-0.5, -1). The problem was in using the degree -1, because it could lead to zero final polarity. Therefore, the current version does not use value "-1" for intensification processing.

Our approach does not need for an intensifier (respectively negation) and the related word to be located as neighbors. They can take any position within one lexical unit, while its length is limited by the value "4" in 4-gram approach. Intensifier and/or negation can be located before or after the related word. Within our approach, not only intensification, but also negation processing is different, based on 4-grams. The paper [8] processes negation by switch negation and shift negation. They were described within the section 2.3. Our approach is different from these negation types. We process all negations found within the whole given contribution together in the second sum of formula (1). Each negation is represented by decreasing of final polarity by two degrees (-2).

3.2 Combining Approach to Using N-grams

The question is how long the length of processed lexical unit should be. We decided to represent the lexical units by n-grams. The value of *n* was fixed on 4 in an experimental way. Shorter value of *n* can separate negation or intensifier from related word. Consequently, we have designed new versions of n-gram approach. The first version has created 4-grams applying shift by one position.

Using 1-grams represents processing each word of a contribution separately and it is a very common technique within the field of opinion analysis. But using 4-grams is not so common. The work [3] uses 4-grams in sentiment detection engine named Umigon for sentiment analysis of tweets. It belongs to lexicon based sentiment classifiers. The text of tweets is decomposed into a list of 4-grams and they are compared with terms in lexicons. Each term in a lexicon is accompanied by a heuristics and a decision rule for 4-gram polarity determination. The work [3] does not present any unified way to polarity calculation as it is used in our approach.

Our approach is based on combination of resulting polarity value obtained by 4-grams and polarity value obtained by 1-grams. The algorithm based on 4-grams is more precise than 1-grams algorithm. The way of combination 1-grams and 4-grams for cases when they disagree is illustrated in Table 4.

Table 4. Combination of 1-grams and 4-grams

1-gram	4-gram	Final classification
positive	neutral	Neutral
negative	neutral	Neutral
neutral	positive	Positive
neutral	negative	Negative
negative	positive	positive or neutral
positive	negative	neutral or negative

The Table 4 represents the way how to determine the result of opinion classification from two classifications: one obtained by 1-grams approach and the second obtained by 4-grams approach. The combining classification can increase the classification precision and it is well known approach in the field of machine learning. The 1-grams approach determinates the words polarity separately. It is more simple technique but has lower precision. The 4-grams approach considers also relations (negation, increasing or decreasing polarity) between four neighboring words and therefore it is more precise. This is a reason, why the final classification is the same as 4-grams classification in the great majority of cases, when the particular classifications of 1-grams and 4-grams disagree. Last two rows of the Table 4 represent the cases, when the particular classifications are strongly opposite.

There is a question, why we have used only 1-grams and 4-grams. The 1-gram usage represents a classical solution of basic problems of opinion analysis. The value N = 4 has been determined experimentally. The 4-grams are just long enough to prevent isolation of a negation or intensification from a related word in the case when related word is not close neighbor of negation (intensification). From this point of view 2-grams and 3-grams can't bring any benefit, because 4-grams can also cover phrases of two or three words.

3.3 Implementation Testing

The designed n-grams approach to opinion extraction was implemented. The implementation was tested on a set of discussion contributions from the portal http://www.mojandroid.sk (discussion thread related to reviews of the mobile telephones HTC One X and HCT One S) and http://www.pocitace.sme.sk (discussion thread related to reviews of two products Asus Transformer Prime TF201 and Asus Transformer Pad TF300T). This first experiment was provided on the set of 42 contributions and 2350 words (average number of words in one contribution was 56).

The implementation was tested also on a set of discussion contributions from the portal http://tech.sme.sk (discussion thread related to reviews of the telephone Samsung Galaxy S4) and from the portal http://www.mojandroid.sk (discussion thread related to reviews of the telephones HTC ONE and Samsung Galaxy S4). This second experiment was done on the set of 71 contributions and 4341 words (average number of words in one contribution was 61).

The resulting precision and recall of our implementation in both experiments can be seen in Table 5. The achieved precision and recall (mainly recall) of negative contributions was quite low. The reason can be lower number of negative contributions than positive contributions in testing data. These results are influenced by contributions containing irony or polarity hidden in the context.

Table 5. Achieved precision and recall of the tests of the designed approach

Experiment	Precision: positive	Precision: negative	Recall: positive	Recall: negative
1.	0.830	0.570	0.652	0.214
2.	0.757	0.419	0.622	0.374

4 Conclusions

The paper introduced an approach based on n-grams focusing on solving the problem of opinion classification to positive or negative polarity. It achieved lower precision within negative contributions while it had quite high precision within positive contributions processing. These results are quite common within the existing opinion classification applications. The lower precision of the classification of negative contributions can be caused by the fact, that majority of web discussions contain less negative contributions.

There is a possibility to upgrade version based on n-grams to achieve higher precision within processing of negative contributions. There is a need to extend the classification dictionary. This version should be enriched by techniques for processing also contributions, which contain only neutral words, but their context is positive or negative. The techniques for processing of irony and ambiguity should be included too. The research in the field of opinion classification has big importance for the future. A successful application of opinion classification can be very helpful in the process of decision making.

Many n-grams can represent noise instead of valuable information. The n-grams, which represent noise, are not a principal problem for our approach from the point of precision. But processing of n-grams with noise could prolong the required processing time. Thus, some filtering and cleaning mechanism could be adopted in the future improvement of our implementation. Another problem of our application is not very fine grained assignment of the constant score to qualifiers. Therefore we consider to modify our application using fuzzy membership functions to handle qualifiers (intensifications).

Acknowledgements. The work presented in this paper was supported by the Slovak Grant Agency of Ministry of Education and Academy of Science of the Slovak Republic within the 1/1147/12 project "Methods for analysis of collaborative processes mediated by information systems".

References

1. Choi, Y., Cardie, C.: Learning with Compositional Semantics as Structural Inference for Subsentential Sentiment Analysis. In: Proc. of the EMNLP 2008, Conference on Empirical Methods in Natural Language Processing, pp. 793–801 (2008)
2. Go, A.: Twitter Sentiment Classification using Distant Supervision. Stanford University, http://cs.stanford.edu/people/alecmgo/papers/TwitterDistantSupervision09.pdf (cit. June 4, 2013)
3. Levallois, C.: Umigon: sentiment analysis for tweets based on lexicons and heuristics, pp. 1–4. Erasmus University Rotterdam, The Netherlands (2013)
4. Liu, B.: Sentiment Analysis and Opinion Mining (Introduction and Survey), pp. 1–168. Morgan & Claypool Publisher (May 2012)
5. Lukáč, G., Butka, P., Mach, M.: Semantically-enhanced Extension of the Discussion Analysis Algorithm in SAKE. In: SAMI 2008, 6th International Symposium on Applied Machine Intelligence and Informatics, Herľany, Slovakia, pp. 241–246 (January 2008)
6. Mach, M., Lukáč, G.: A Dedicated Information Collection as an Interface to Newsgroup Discussions. In: IIS 2007 - 18th International Conference on Information and Intelligent Systems, Varazdin, Croatia, September 12-14, pp. 163–169 (2007) ISBN 978-953-6071-30-2
7. Smatana, M., Koncz, P., Paralič, J.: Semi-automatic Annotation Tool for Aspect-based Sentiment Analysis. FEI Technical University of Kosice, pp. 1–3 (2013)
8. Taboada, M., Brooke, J., Tofiloski, M., Voll, K., Stede, M.: Lexicon-Based Methods for Sentiment Analysis. Computational Linguistics 37(2), 267–307 (2011)
9. Thelwall, M., Buckley, K., Paltoglou, G., Cai, D., Kappas, A.: Sentiment Strength Detection in Short Informal Text. Journal of the American Society for Information Science and Technology 61(12), 2544–2558 (2010)

Towards a Cloud-Based Group Decision Support System

Ciprian Radu, Ciprian Cândea, Gabriela Cândea, and Constantin B. Zamfirescu

Ropardo S.R.L., Sibiu, Romania
{ciprian.radu,ciprian.candea,gabriela.candea}@ropardo.ro,
zbc@acm.org

Abstract. Decision Support Systems have constantly benefited from the technological advances in Computer Science. Cloud Computing is a technology that could become useful for the Decision Support Systems, too. Moreover, Decision Support Systems may support the "global brain" programming paradigm. This paper presents iDS, a system designed to become a collaborative- and cloud-based Group Decision Support System. This Decision Support System will be made available as a Business as a Service model. After presenting the architecture of the developed iDS, the main design directions for integrating it into the cloud are described. The goal is to obtain a platform that supports collective intelligence, in terms of human-computer networking.

Keywords: Group Decision Support System, Cloud Computing, Collaborative Platform, Collective Intelligence.

1 Introduction

The Group Decision Support System (Group DSS, GDSS) is defined as a combination of communication, decision and computer technologies working together to offer support for decision phases like: problem identification, formulation and solution generation, during group meetings [1].

Nowadays, the results of collaboration activities inside a group are based on coordination, decision making and negotiation. Most of these phases rely on knowledge of the organization. In many organizations, decision making situations are naturally recurring and the majority of them are critical.

Current research at Ropardo results in a collaborative decision making support platform, named iDS (iDecisionSupport) and designed to be easy to use by the majority of users (thus avoiding long trainings and preventing user rejection) [2]. The implemented decisional model is based on the Shared Plans theory [3]. It was tested for the first time in a software prototype by our group in 2001 [4].

Cloud Computing is not something new because it essentially refers to computing as a utility [5]. However, the term "cloud" started to become popular in 2006 [6].

According to [7], the essential characteristics of Cloud Computing are:

on-demand self-service, broad network access, resource pooling, rapid elasticity, measured service.

A. Bădică et al. (eds.), *Recent Developments in Computational Collective Intelligence*,
Studies in Computational Intelligence 513,
DOI: 10.1007/978-3-319-01787-7_18, © Springer International Publishing Switzerland 2014

Cloud Computing technologies are mainly classified in terms of service and deployment models [7]. There are three types of service models. With the Software as a Service (SaaS) model, the cloud offers applications that are available to clients through interfaces like web browsers. The user has no control over the cloud infrastructure. With Platform as a Service (PaaS) the consumer cannot control the cloud infrastructure but, custom applications (developed or acquired by the service user) can be deployed into the cloud. In Infrastructure as a Service (IaaS) the service user still cannot control the cloud infrastructure but, access to operating systems, storage and even some limited networking control are allowed. From the service models point of view, everything is a service. Hence, the XaaS term was coined [5], [8].

A Cloud Architecture has several deployment models: private, community, public and hybrid cloud. A private cloud is used solely by a single organization. The community cloud is meant for a certain community of users, from multiple organizations, which have some common interests. The public cloud is available to anybody. A combination of at least two of the above three types of clouds forms a hybrid cloud.

In this paper we focus on the advantages of the Business as a Service (BaaS) model, in which the system is not simply provided as a service to the customer but, it is also managed so that the business goals are met. Business as a Service proposes to the companies a mix between business support, proven methodology, strategy support and hosted technology. The BaaS model is cost-effective and it provides better business results because customers can benefit from the vendor's experience in how to best use the delivered system.

The main contribution is a collaborative- and cloud-based GDSS at BaaS level. The goal is to obtain a platform that supports human-computer networks.

The remainder of this paper is organized as follows. Section 2 briefly presents the related work. Section 3 describes iDS' architecture. Section 4 concludes this paper and outlines some further work directions.

2 Related Work

Besides supporting information access [9], a GDSS can, at the same time, radically change the dynamics of group interactions by improving communication, by structuring and focusing problem solving efforts and by establishing and maintaining an alignment between personal and group goals.

Traditional decision making processes follow an iterative process, having the following phases: analysis and definition of the problem, divergence and then convergence on the set of possible solutions and lastly choosing the final one. Since the first DSSs [1], many improvements were done by the research community, based on technological advances. For example, Web 2.0 and 3.0 technologies are nowadays used to support collaborative-based GDSSs. Focus is put on the social aspects by adopting social network models into DSSs [10].

This paper focuses on the integration of a GDSS solution with Cloud Computing technology. According to [6], the current Cloud Computing state-of-the-art implementations have the following architectural characteristics: (1) uniform high capacity (maximum communication bandwidth between any two computers from the data center), (2) free Virtual Machine migration (a Virtual Machine can migrate from a physical machine to another one rapidly and easily), (3) fault-tolerance, (4) scalability and (5) backward compatibility.

DSSs using collaborative- and cloud-based platforms represent a rather new research field. GRUPO-MOD [11] is a collaborative decision making system based on analytical hierarchy processes. It is implemented using a client – server web architecture, which facilitates asynchronous decisions. Its authors stress out that modern collaborative decision making systems are transiting to Cloud Computing enabled solutions. To this end, they intend to bring their system at the SaaS level.

The iDS system presented in this paper is a collaborative decision making system based on a client – server web architecture, similar with GRUPO-MOD. Both systems adopt a plugin-based approach. The main difference is that we intend to bring iDS to an upper level: BaaS. While at SaaS level the vendor is responsible for the software applications provided, at BaaS level, the vendor is responsible for the entire business of a company, i.e. how and in what processes are the software applications used. This way, a company may remain focused on how to develop its business rather than how to implement specific actions.

Moreover, we intend to obtain with iDS a platform that supports the “global brain” concept, i.e. the network of all people and computers from our planet [12].

3 The iDS Architecture

This section presents the architecture of the developed iDS system. We start by presenting its decision making process model. Then we focus on its main subsystems. Finally, we present further architectural aspects intended for making iDS a computing utility (by integration with Cloud Computing technologies).

iDS is designed as a generic framework for decision support. To better detail its architecture, we will give an example of how it may be used for meeting management.

3.1 The Decision Making Process Model

Decision making can be defined as the process of choosing the best solution, from two or more possible ones, in order to solve a particular problem. It is especially characteristic to groups of people or to organizations.

The decision making model adopted in iDS is hierarchical. The decisional activities are associated with projects. A project has one or more plans. Any plan can have sub-plans and both are time-boxed. Otherwise, it can have one or more sessions. Fig. 1 presents an example of a decision making process, which can be obtained with such a model. Project A, has two plans (i.e. A and B), each of them showing two key characteristics of this model. Firstly, a (sub-)plan may have several sessions that can run, in time, sequentially or in parallel. Secondly, sessions can be further organized using sub-plans. It is not necessary that each leaf (sub-)plan (leaf because the model is a tree) has at least a session. This is because the model may evolve in time. A session is defined as the period of time allocated to a specific decision making activity.

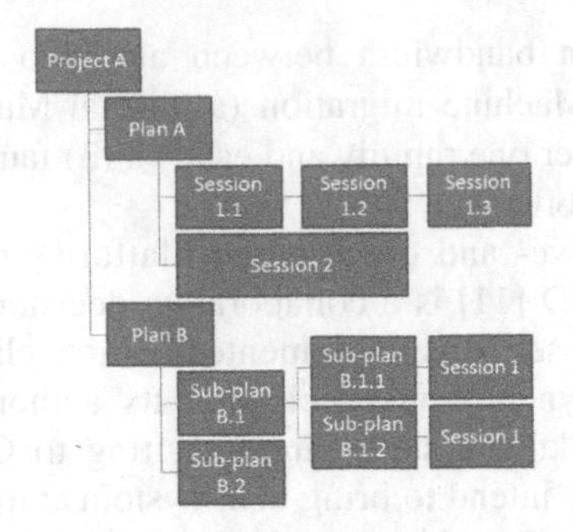

Fig. 1. iDS decision making process example

The session is modeled in iDS as a sequence of well-defined phases: *draft*, *commit*, *work* and *report*. While in *draft*, only the session's author has access to it. This phase is used for configuration purposes, which may also be done during *commit*. The difference between *draft* and *commit* is that once *draft* ends, the session becomes public to anyone invited to participate. Each person can accept or deny the participation to the session until the end of the *commit* phase. After the *commit* phase, the session enters in the *work* phase. There can be a break between these two phases, which allows for the work to be scheduled at a later time. During *work*, the session participants (which accepted to attend the session), use a decision support tool in order to make some decisions. The results of their work will be available after this phase ends, i.e. in the *report* phase.

This decision making process model may be enhanced with collective experience. For example, collaborative deliberation [12] is a "global brain" programming metaphor that involves both computers and humans in obtaining, analyzing and selecting solutions. Decision processes are part of the "global brain" programming language.

For the meeting management case study, hierarchical plans may be used to define a project's objectives. Each objective may have several sub-plans, which may represent different meetings (required for reaching the objective). Each sub-plan may have several sessions, which form the meeting's agenda.

3.2 The Developed iDS System

iDS operates as a software system accessible through the Internet. As such, both synchronous and asynchronous decision processes are supported. It provides a range of services that stimulate the collaboration between individuals. Using a common Application Programming Interface (API), iDS allows for pluggable decision support tools. This system was already successfully integrated in a virtual factory environment [13].

The system's core is the decision support engine. It provides various iDS services over a default Graphical User Interface (GUI). The decisional support is available through different decision support tools. The tools can be integrated with the iDS core through the tools plugin support module that acts as an iDS Tools Connector. The following subsections provide more details about the architecture of the developed iDS system. As it will be shown throughout the paper, iDS is designed to become a computing utility, by integrating it into the cloud.

The iDS Server

The iDS server is the core of the system. Fig. 2 a) presents a top view on the server's architecture by showing its main subsystems and modules.

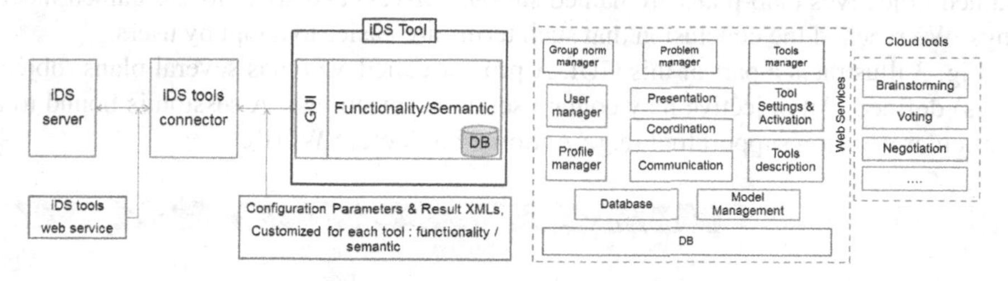

Fig. 2. a) The iDS server architecture b) iDS server components and cloud tools

iDS uses a Relational Database Management System for data persistence but, we are also investigating the benefits of using a semantic data model [13].

To enable the cloud paradigm, the system implements a decentralized architecture. It presents the concept of server and independent tools as a service. The system's communication mechanism is based on syntax (message format) and semantics (message meaning). For the moment, the system is not implementing yet a communication standard (like KQML) but, uses XML and XSLT for encoding and interpreting the messages. iDS considers the fact that, for a successful decision support, different tools must be used and nowadays many of them can be accessible from the cloud. For this reason, the iDS server is as in Fig. 2 b), where different layers that support the users and their decisional process are presented.

iDS exposes its functionalities through web services. Data access is obviously secured. The entire iDS server can be administered remotely using a web application.

The iDS server has three pillars: groups, decisions/problem and supporting tools, and adopts the collective intelligence principles (openness, peering, sharing and acting globally).

The Group pillar refers to the iDS users, their roles and profiles within the system. Its main objective is to offer collaborative functionalities.

The Decision/problem pillar essentially refers to the decision making process model. This subsystem has the objective of managing the lifecycle of each session automatically, while structuring sessions hierarchically. It also communicates with the decision support tools by configuring them, commanding them and gathering their output. A reporting subsystem facilitates the aggregation of data received from the tools.

The decision support tools form another important pillar, which will be described in a later section of the paper.

To follow-up on the meeting management example, groups of people may be defined for the considered project (using the administration system). They may be involved in the decisions of different objectives, having different responsibilities (roles). Some may coordinate the project at a global level, setting its objectives. Others may act as decision making facilitators [14].

The Default Graphical User Interface

The default iDS GUI was developed as a Web 2.0 application. Users can experiment with the decision making process model available in iDS. For convenience, plans are named objectives (sub-plans are named sub-objectives) and sessions are named meetings. We reached the conclusion that such terms are easier to adopt by users.

Fig. 3 illustrates a part of this GUI. A project called VFF has several plans (objectives) defined. An objective may contain sessions (meetings). A session is bound to a particular decision support tool (e.g.: Action Plan, Vote, SWOT).

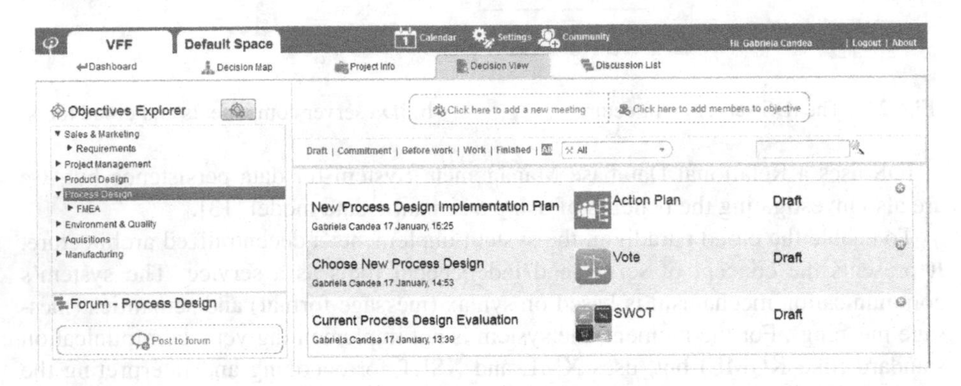

Fig. 3. The iDS default GUI

Other functionalities provided by this GUI may be observed. The user has a profile, access to a personal calendar and may exchange ideas and share information by having access to forums and discussion lists in key areas of the application. Overall, the iDS system provides functionalities specific to Collaborative Platforms.

This default GUI presents an implicit and generic view of the iDS system in general. It is intended however to provide particular views as well. A particular view will correspond to a specific business model. Such approach will allow iDS to be customized for different businesses. As it will be showed later in the paper, Cloud Computing provides advantages for a BaaS approach.

Meeting management will have a particular GUI, focused on objectives, meetings and their agenda. The GUI will access the iDS system's web services in a secure manner, starting from user authentication and continuing with defining objectives, meetings and configuring decision support tools.

Tools in Cloud

The iDS Tools Connector (ITC) allows any software tool to be connected and communicate with the iDS system. It essentially establishes a communication protocol between iDS and the software tools, based on an API for the iDS system and an API for the tools. Potentially, any software can be adaptable to iDS.

The iDS API is a web service that allows the iDS server to: instantiate, disembody and command a tool. The tool is allowed to be in three sequential phases: configuration, run and report. A tool is initially configured manually by the user, or automatically by the system (as in Fig. 2 a)). Finally, the output it produces is made available to the iDS system. An ontology was defined for the interaction and communication between any tool and the iDS server. Based on the ontology that is represented in Fig. 4, we ensure that all tools available the cloud will refer the terms and semantics of messages in a similar way [15].

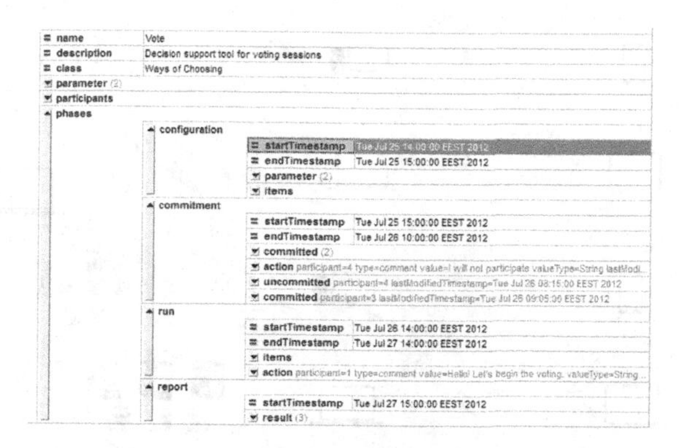

Fig. 4. Tool message structure

The tool API has two parts. Firstly, there is a programming interface that contains a list of operations through which the ITC can request the tool to perform actions. The requests coming from iDS are delegated via this interface. Secondly, there is a web service through which the tool can send data to the iDS server, asynchronously.

We successfully integrated so far decision support tools like: vote, brainstorming, SWOT, action plan, Mind Map and categorizer. As it will be showed in a subsequent section of the paper, Cloud Computing can increase the advantages of our approach.

For meeting management, facilitators will request iDS different support tools. As such, tools will be instantiated and then the users may configure and use them. When a tool becomes unneeded, iDS will remove it and free the resources allocated for it. In case a customer needs a decision support tool which is not available in iDS, such tool could be developed and easily integrated in the iDS.

3.3 Further Architectural Considerations

The architectural design of the iDS system includes its integration with Cloud Computing technologies. The objective is to create from iDS a GDSS which benefits from some of the characteristics of Cloud Computing. The iDS computing capabilities are foreseen to be provided without human interaction, through an on-demand self-service, and also to be available over the Internet, on different platforms like mobile phones, tablets, desktop and laptop computers (broad network access). Each iDS user will benefit from rapid elasticity of the computing capabilities. This will allow each user to quickly scale those iDS system parts which are mostly needed. The iDS resources will be made available (through virtualization) to multiple clients, without knowing their express (physical) location. Resource pooling will be thus another Cloud Computing characteristic that will be inherited by iDS.

The iDS system will be easily and broadly accessible through the network. It will be distributed across four levels: BaaS, SaaS, PaaS and IaaS.

Different customers will be able to manage their businesses' decision making processes by interacting with the iDS BaaS level, which is detailed in Fig. 5 a).

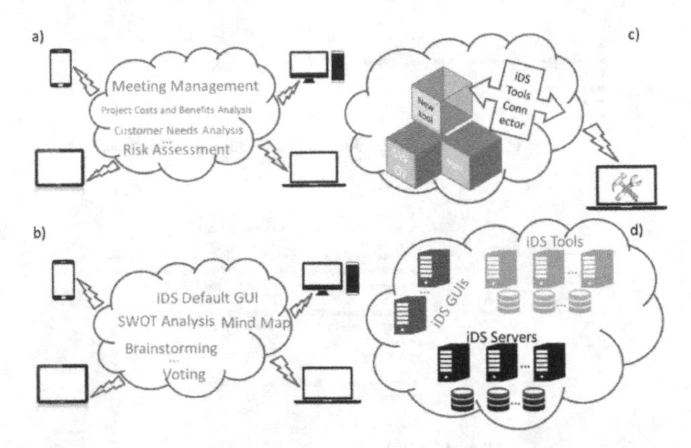

Fig. 5. a) iDS BaaS level; b) iDS SaaS level; c) iDS PaaS level; d) iDS IaaS level

At BaaS level, iDS will allow users to manage different types of decision processes that are encountered in their businesses. The generic decision process model developed in iDS can be used for example by a company to: manage its meetings, or to perform analysis regarding customer needs or project's costs and benefits, or even for risk assessment. Other scenarios can obviously be considered. In order to meet the particular needs of a certain business, different views of the iDS default GUI will be developed. They will then be integrated in a web portal. Therefore, each business will take advantage only from the iDS decision making functionalities that are of interest for it, presented in a customized manner.

As it is presented in Fig. 5 b), at the SaaS level, iDS will provide all its software modules (default GUI, decision support tools) that can offer services to the end user.

The main difference between the iDS BaaS and SaaS levels is that, at business level, iDS is presented in a customized form, which can vary from customer to customer.

At the SaaS level, the user may be interested in experimenting with the iDS decision process model, by using the iDS default GUI. Also, the available decision support tools may be tested. To this end, the current iDS implementation already provides a public (default) space that can be used by users for trying out iDS.

At PaaS level (Fig. 5 c)), new tools may be plugged in the iDS system by exposing to the developer well defined APIs. A thorough presentation of the iDS Tools Connector is considered beyond the scope of this paper.

Finally, the iDS IaaS level is presented in Fig. 5 d). This level will only be accessible to the administrators of the iDS Cloud architecture.

An important problem will be addressed at this level: what deployment models of the iDS system (with all its components) must be used in order to properly meet the customers' needs and expectations. Since customers will most likely manifest particular interests regarding the iDS system, which are driven by their businesses, resourcc allocation will have to differ from one case to another.

One important objective for Cloud Computing is to require the customer to pay only for the needed resources, when they are needed. To this end, for example, decision processes should occupy only the computational resources (tools) that they need, when they need them. Moreover, each client will require computational resources only for the decision support tools that he or she needs.

At BaaS level, meeting management is just one example (see Fig. 5 a)) of how the iDS services may be aggregated into a business level service. In the cloud, a middle layer will be responsible with provisioning iDS services for different organizations and their business. For example, the relevant decision support tools will be made available from the most suitable geographic locations. Another example is that only the GUIs required for an organization's business will have resources allocated. Such custom, business-oriented configurations will be made in a semi-automatic way, without concerning the customers with the implementation details.

4 Conclusions and Further Work

This paper presented the developed iDS solution, a collaborative group decision support system. Its main architectural aspects were described where one important novelty was presented: the ontology for data exchange between tools. iDS provides a flexible decision making process model and it has pluggable decision support tools. Based on tools that now can run in the cloud (to support different business scenarios), meeting management is a wide spread scenario that is met in any company, and that can be easily supported by iDS.

To introduce decisional processes into the cloud, several actions, which need to be taken at infrastructure, platform, service and business level, were presented. However, there are still many research challenges in the field of Cloud Computing [5], [6], which will have to be considered when integrating iDS into the cloud.

Different advantages were presented for each level that is available in the cloud paradigm, and the ultimate goal, BaaS, was evaluated. At the BaaS level any company can benefit from the best in class tools as well best in class business models to manage their problems.

We will also study the integration of iDS with the CowdLang platform [16] towards collective intelligence and programming the "global brain". Our idea is to use iDS for the decision operators from the CrowdLang programming language.

References

1. DeSanctis, G., Galluoe, B.: A Foundation for the study of Group Decision Support Systems. Management Science, 589–609 (1987)
2. Georgescu, V., Candea, C., Zamfirescu, C.B.: iGDSS - Software Framework for Group Decision Support Systems. In: Proceedings of The Good, The Bad and the Unexpected Conference (2007)
3. Grosz, B., Kraus, S.: Collaborative plans for complex group action. Artificial Intelligence 86, 269–357 (1996)
4. Zamfirescu, C.B., Candea, C., Luca, S.I.: On Integrating Agents Into GDSS. In: Filip, F.G., Dumitrache, I., Iliescu, S.S. (eds.) Preprints of the 9th IFAC/IFORS/IMACS/IFIP/Symposium on Large Scale Systems: Theory and Applications, Bucharest, Romania, pp. 231–236. ICI Press (2001) ISBN 973-98407-8-7
5. Armbrust, M., et al.: Above the Clouds: A Berkeley View of Cloud Computing. EECS Department, University of California, Berkeley (2009)
6. Zhang, Q., Cheng, L., Boutaba, R.: Cloud computing: state-of-the-art and research challenges. Journal of Internet Services and Applications 1(1), 7–18 (2010)
7. Mell, P., Grance, T.: The NIST definition of cloud computing. NIST Special Publication 800:145 (2011)
8. Rimal, B.P., Choi, E., Lumb, I.: A taxonomy and survey of cloud computing systems. In: Fifth International Joint Conference on INC, IMS and IDC. In: NCM 2009, pp. 44–51 (2009)
9. DeLone, W.H., McLean, E.R.: Information systems success: The quest for the dependent variable. Information Systems Research 3(1), 60–95 (1992)
10. Antunes, F., Costa, J.P.: Integrating Decision Support and Social Networks. In: Advances in Human-Computer Interaction (2012)
11. Thimm, H.: Cloud-Based Collaborative Decision Making: Design Considerations and Architecture of the GRUPO-MOD System. IJDSST 4(4), 39–59 (2012)
12. Bernstein, A., Klein, M., Malone, T.W.: Programming the global brain. Commun. ACM 55(5), 41–43 (2012)
13. Candea, G., Candea, C., Radu, C., Terkaj, W., Sacco, M., Suciu, O.: A practical use of the Virtual Factory Framework. In: 14th International Conference on Modern Information Technology in the Innovation Process of the Industrial Enterprises, Budapest, Hungary (2012)
14. Dickson, G., Poole, S., DeSanctis, G.: An overview of the GDSS research project and the SAMM system. In: Bostrom, Watson, Kinney (eds.) Computer Augmented Teamwork: A Guided Tour. Van Nostrand Reinhold (1992)
15. Jasper, R., Uschold, M.: A Framework for Understanding and Classifying Ontology Applications. In: Twelfth Workshop on Knowledge Acquisition Modeling and Management, KAW 1999 (1999)
16. Minder, P., Bernstein, A.: Crowdlang-first steps towards programmable human computers for general computation. In: Proceedings of the 3rd Human Computation Workshop (HCOMP 2011). AAAI Press (2011)

MapReduce-Based Implementation of a Rule System

Ryunosuke Maeda[1,*], Naoki Ohta[2], and Kazuhiro Kuwabara[2]

[1] Graduate School of Science and Engineering, Ritsumeikan University
[2] College of Information Science and Engineering, Ritsumeikan University,
1-1-1 Noji-Higashi, Kusatsu, Shiga 525-8577 Japan

Abstract. As information and communication technologies advance, large amounts of data are created everyday. The demands for processing such big data are also increasing. To meet them, the MapReduce framework has been proposed and is now widely used. On the other hand, a rule-based system is used to implement such an intelligent system as an expert system. For applying a rule-based system to process large amounts of data, we propose a method that implements a rule system based on the MapReduce framework. We constructed a simple rule system using Hadoop, which is an open source implementation of the MapReduce framework, and compared several methods of executing a rule system. Our experimental results indicate the potential of a rule system implemented using the MapReduce framework.

Keywords: rule-based system, MapReduce, Hadoop, inference.

1 Introduction

The advance of information and communication technologies constantly creates large amounts of data. The need for processing such large amounts of data is also increasing. To process such big data, a distributed system comprised of a cluster of computers is often used. One such system is the MapReduce framework [4], whose open source implementation, Apache Hadoop [1], is used in many places. A MapReduce program decomposes a given job into *map* and *reduce* tasks. Since these tasks are distributed automatically by the MapReduce framework, a program can easily exploit the underlying distributed computers.

On the other hand, as a model of an intelligent system such as an expert system, a rule-based system is widely used. The format for exchanging rules written in a variety of rule systems was created as Rule Interchange Format (RIF), which is one of the W3C recommendations [14]. RIF defines not only the syntax of the rules but also their semantics [8]. More specifically, RIF defines the core part of the rules and their extensions, which consist of an action-based rule system (production system) and a logic-based rule system. The rule system is also applied in distributed systems [2]. It is preferable that rule-based systems

* Currently with The Japan Research Institute, Limited.

A. Bădică et al. (eds.), *Recent Developments in Computational Collective Intelligence*, Studies in Computational Intelligence 513,
DOI: 10.1007/978-3-319-01787-7_19,

can process large amounts of data with a cluster of computers in a distributed fashion.

Therefore, we propose methods that implement a rule-based system using the MapReduce framework with which we expect that large amounts of data can be easily processed by a rule-based system. We implemented our proposed methods using Hadoop and experimentally evaluated them.

The rest of this paper is structured as follows. The next section briefly describes related research. Section 3 describes our rule system implemented with the MapReduce framework. Section 4 describes our experiments and discusses their results. The final section concludes the paper.

2 Related Works

The MapReduce-based rule matching algorithm was proposed for a production system [3]. This algorithm is based on the Rete algorithm [5], a well-known efficient pattern matching algorithm for production systems. In this MapReduce-based algorithm, a master node analyzes the contents of the rules and assigns map and reduce tasks to a slave node to execute on a cluster of computers. However, it does not exploit a distributed file system that is often used with MapReduce-based algorithms, and it remains unclear how much scalability can be obtained. In contrast to this approach, we assume Hadoop implementation of the MapReduce framework and place the data on an accompanying distributed file system called the Hadoop Distributed File System (HDFS). The assignment of each task to a node is done by Hadoop itself to easily achieve scalability.

From the viewpoint of a reasoning system, the MapReduce framework was used in the inference system of the Resource Description Framework (RDF), which constitutes the basis of the Semantic Web [13]. In this inference system, for example, the MapReduce framework derives new RDF triples from a large set of them from the given rules of the RDF Schema (RDFS). Since the rules that derive the facts (triples) in RDFS are given, this system analyzes the rules beforehand and creates specialized programs to reduce the bottleneck in the execution. In contrast, this paper focuses on a more general rule system where rules are basically defined by a user program.

In addition, the MapReduce framework is applied to processing a large amount of RDF data [6, 9–11]. For example, [9] proposed a method that uses MapReduce for pattern matching in the WHERE clause of SPARQL, which is a query language for RDF. In SPARQL graph pattern matching, each pattern in the WHERE clause matches a triple in the RDF dataset. Graph pattern matching in the WHERE clause of SPARQL resembles the matching antecedents of the rules. Their ideas can probably be applied to the optimization of rule execution. However, in the case of a rule system, after a rule is executed, new facts are derived and another matching needs to be initiated. In this sense, a MapReduce-based rule system needs an additional mechanism to handle iterations in its rule execution.

3 MapReduce-Based Implementation of a Rule System

3.1 Example Rule System

To investigate MapReduce-based implementation, we constructed a simple rule system that is not intended to be a full-fledged system; it only includes a basic function. The system consists of facts and rules. A fact is assumed to be a triple modeled after an RDF triple. A rule consists of an IF part and a THEN part. The IF part consists of condition patterns, each of which matches a fact. The THEN part also consists of patterns, each of which specifies a newly derived fact when the IF part of the rule is satisfied. This rule system, which is a declarative type, is not a production system. Thus, the THEN part does not contain an operation to modify an existing fact, and facts are monotonously increased as the rules are fired.

The rule system is executed as follows. The system searches for a rule whose IF part is satisfied. If such a rule is found, the facts that are specified in its THEN part are added to the system. If the same fact is derived with an existing fact, it is simply ignored. The execution of the rule system ends when no more new facts can be derived. Since the rule system is not a production system type, we can safely disregard the synchronization problem [7] in parallel rule firing in a production system. Since the rules can be executed in parallel, the key issue in the MapReduce-based implementation is how to speed up the matching between rules and facts.

3.2 MapReduce Programming Model

In the MapReduce framework, a program defines the map and reduce tasks. The map task outputs a key-value pair. The outputs of the map tasks are collected and sorted by keys. For each key, a reduce task is invoked, and the outputs of the map tasks are distributed to the reduce tasks. In the following, we explain our proposed implementation methods assuming that Hadoop is used.

3.3 Baseline Implementation: Version 0

First, we consider a naive implementation of the rule system on top of Hadoop. Since the number of facts usually exceeds the number of rules, we consider how the facts should be distributed. We also assume that each task can hold all the rule definitions because the number of rules is assumed to be small. This assumption is reasonable since facts may be created automatically, but rules are basically written by hand.

The facts are stored in the Hadoop Distributed File System (HDFS). The rules, which are pre-compiled into Java data structure, are included in the JAR file of the rule system program that are distributed to each node.

Each map task receives part of the facts and performs pattern matching with each condition of the IF part of the rules. The map task's output is the key-value pairs where the key is a *rule-id* that specifies the rule, and the value is

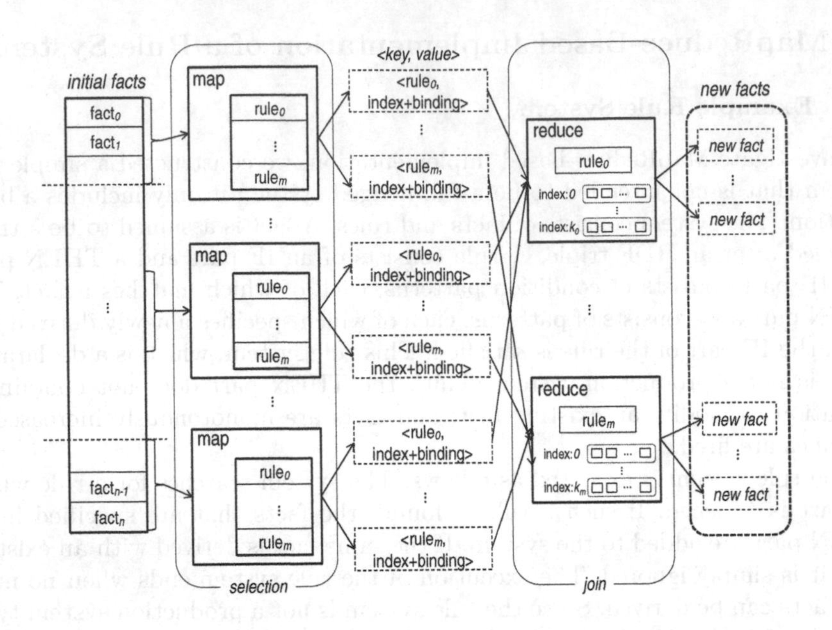

Fig. 1. Map and reduce tasks (*fact producing phase*)

a concatenation of the index of the matched condition in the corresponding rule and the binding information of the variables that appear in the matched condition.

Since the output's key of the map task is a rule-id, a reduce task is basically created for each rule. When the IF part of a rule has multiple conditions, the reduce task effectively calculates the *join* of the matched results for each condition in the IF part. When all the conditions are satisfied, new facts are created that are specified in the THEN part of the rule. To distinguish a newly created fact from the existing ones, each fact is tagged with a number that corresponds to the loop count of the rule execution. Figure 1 shows the map and reduce tasks in the MapReduce job, which we call the *fact producing phase*. As shown in this figure, a map task performs a kind of *selection* operation in the sense that it produces partial results selected from all the facts to be fed into a reduce task that performs a *join* operation.

Since different rules may produce the same fact, we need to remove duplicate facts. This is done by running another MapReduce job which we call the *fact cleansing phase*. In this phase, both the newly created facts and existing ones are passed to a map task where a hash key is calculated for each fact. The generated hash key is used as the output's key of the map task. Its value is a fact itself. Each reduce task receives facts with the same hash key, and removes the duplicate facts where the older fact has priority over a newer one.

The number of facts created is counted using a counter function of Hadoop. In this way, the master node can easily determine whether a new fact is created by examining the counter's value. If a new fact is created, another *fact producing*

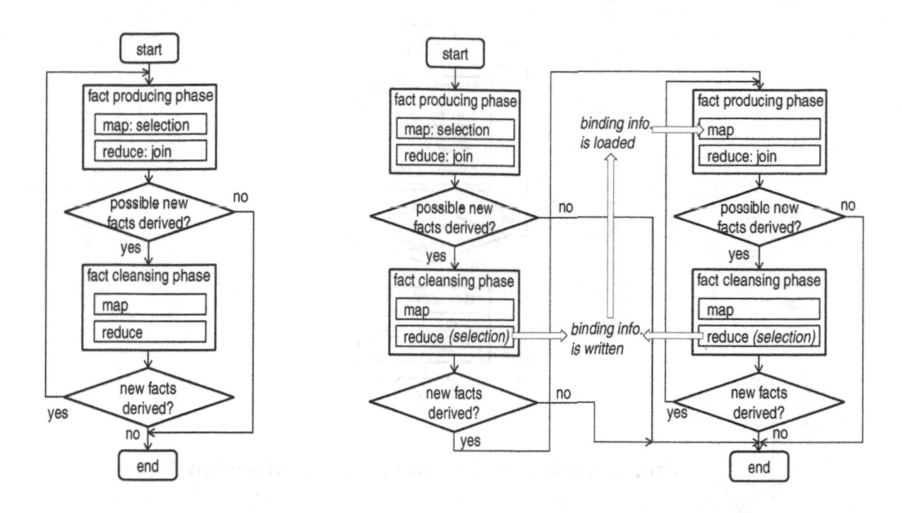

Fig. 2. Version 0: baseline implementation

Fig. 3. Version 1: storing binding information

phase is started. This loop continues until no more facts are created. Figure 2 shows an overview of the baseline implementation, which we call *version 0*. In the following, we consider two speed-up techniques: one reduces the number of pattern matchings, and the other reduces the number of iterations of MapReduce jobs.

3.4 Storing Binding Information: Version 1

In the baseline implementation, for each *fact producing phase*, facts and rules are matched repeatedly. One possible optimization is to store the matching results on HDFS. During subsequent iterations, the stored results are used instead of performing pattern matching facts with rules.

In this algorithm (called *version 1*), the following modifications are made to the baseline implementation (*version 0*). In the first *fact producing phase*, a reduce task outputs the binding information of the initial facts to the HDFS. In addition, in the *fact cleansing phase*, a reduce task also matches the newly created facts with the IF part of the rules and writes them to HDFS. Since Hadoop allows multiple key-value pairs to be written, both the facts and binding information can be outputted by the same task.

In subsequent rule execution iterations, a map task simply retrieves the binding information from the HDFS and outputs it with a key of the rule-id without the pattern matching of facts and conditions in the IF part of the rules (Fig. 3).

3.5 Eliminating MapReduce Jobs: Version 2

In the baseline implementation (*version 0*), for each cycle of rule execution, duplicate facts are checked and removed by invoking another MapReduce job

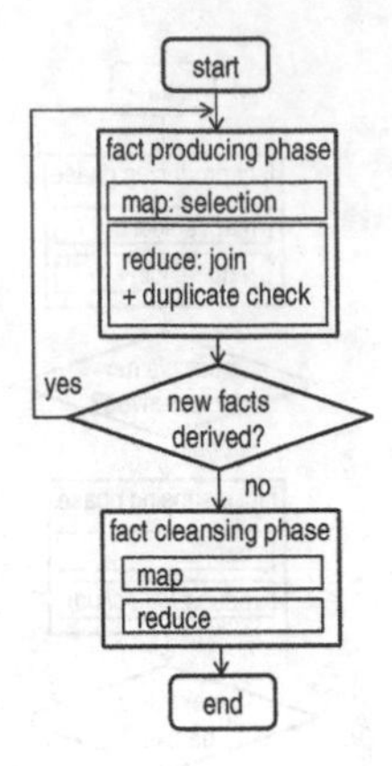

Fig. 4. Version 2: eliminating MapReduce jobs

(*fact cleansing phase*). Since a MapReduce job incurs high overhead, the number of MapReduce jobs should be reduced as much as possible.

In this method (*version 2*), instead of checking duplicate facts by invoking another MapReduce job, they are removed at the reduce task of each *fact producing phase*. Since it is possible that identical facts are derived by different rules, redundant matching may occur in the next iteration, even if a reduce task checks duplicate new facts at each *fact producing phase*. However, no redundant facts are created from the same rule (since a reduce task corresponds to one rule, the same facts produced from the same rule can be ignored), and duplicate facts are not propagated to further iterations. When no more new facts are derived, the duplicate facts are checked by invoking a MapReduce job as in the baseline implementation (Fig. 4).

3.6 Combination of Two Optimization Methods: Version 3

Since the aforementioned two speed-up methods are mutually independent, we consider an algorithm that incorporates both. More specifically, in this version of the algorithm called *version 3*, a reduce task both stores the binding information in the HDFS and checks the duplicate facts.

4 Experiments

4.1 Experiment Settings

To evaluate our proposed algorithms, we conducted experiments with a cluster of 14 computers (nodes). Each node's specifications are shown in Table 1. One node is used as a master node and the others as slave nodes. We experimented with four configurations with different numbers of slave nodes: 2, 3, 6, and 13.

Table 1. Hardware and software of each node

CPU	AMD Turion II Neo N40L 1.5GHz
Memory	DDR3-1333 8GB
HDD	SATA 7200rpm 500GB
Network card	NC107i PCI-E Gigabit
OS	CentOS5
Hadoop	Cloudera CDH3u5

```
[rule01:
  (?x father_of ?y)(?y father_of ?z) -> (?x grandfather_of ?z)]
[rule02:
  (?x grandfather_of ?y)(?y father_of ?z) -> (?x great_grandfather_of ?z)]
[rule03:
  (?x great_grandfather_of ?y) -> (?x ancestor_of ?y)]
```

Fig. 5. Rules used in experiments (partial)

4.2 Rules and Facts

Some of the rules used in the experiments are shown in Fig. 5. The syntax of the rules is loosely based on the rule engine of a Semantic Web toolkit called Apache Jena[1]. For example, `rule01` means that a father of a father is a grandfather. In addition to the rules shown in Fig. 5, we added rules that replaced `father` with `father1` and `father2`. Rules that replaced `father` with `mother` were also added. Thus, the total number of rules was 18 ($= 3 \times 3 \times 2$).

Figure 6 shows the example facts used in our experiments. The facts are created in a way that part of a fact that matches a variable in a rule is sequentially numbered. We created three different sets of facts. Each set consists of 96,000, 960,000, and 2,400,000 initial facts.

```
(A1 father_of B1)
(A2 father_of B2)
...
(C1 mother_of D1)
(C2 mother_of D2)
...
```

Fig. 6. Facts used in the experiments (partial)

4.3 Results

We ran the rule system three times for each experimental setting and calculated the average execution times. Figures 7a, 7b, and 7c show the execution times of four versions of the algorithms with different numbers of initial facts. In these graphs, the horizontal axis indicates the number of slave nodes used.

[1] `https://jena.apache.org/documentation/inference/`

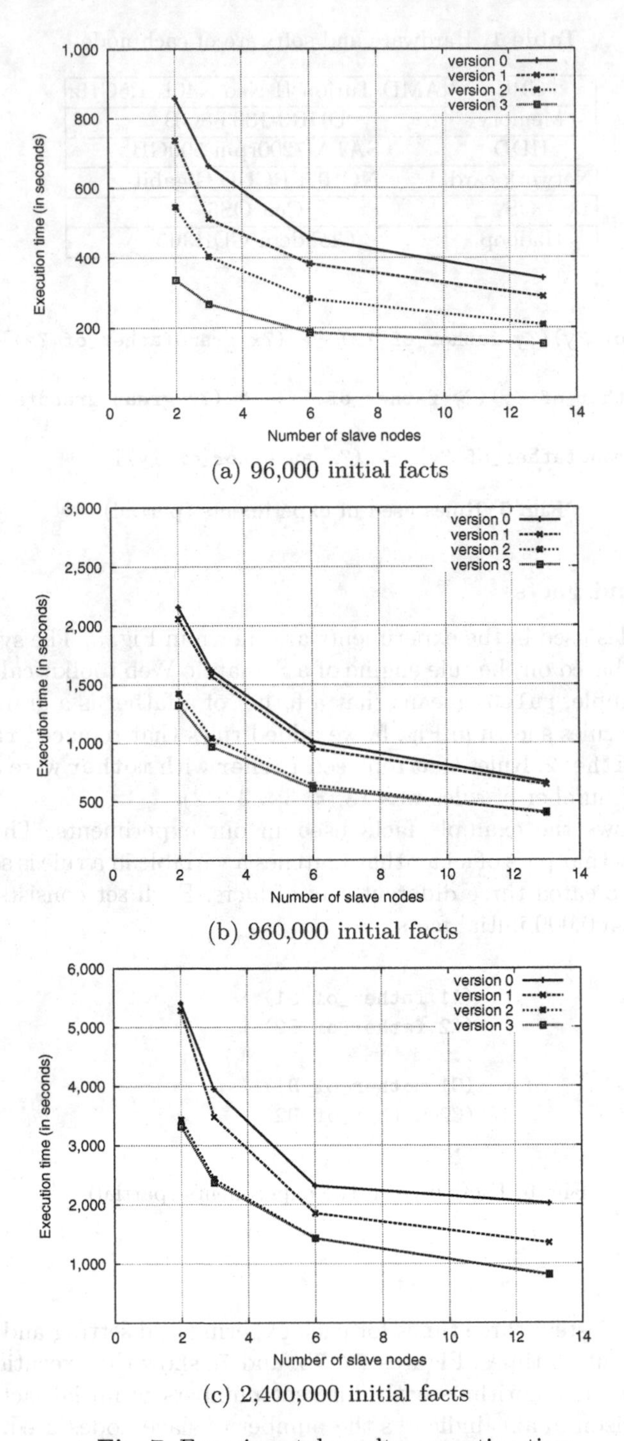

(a) 96,000 initial facts

(b) 960,000 initial facts

(c) 2,400,000 initial facts

Fig. 7. Experimental results: execution time

4.4 Discussion

As seen in our experimental results, by increasing the number of slave nodes, we shortened the execution times. Thus, our proposed algorithms effectively exploited the underlying distributed computers.

By comparing the results of versions 0 and 2 and versions 1 and 3, our method of reducing the number of MapReduce jobs is quite effective, because the overhead in invoking MapReduce jobs is large.

Our method that reduces pattern matching by storing the intermediate results (binding information) is also clearly effective when the number of facts is small (Fig. 7a). However, when the number of facts is increased, the difference between them is not so large (Figs. 7b and 7c). When the number of facts is large, facts tend to be distributed more widely across the slave nodes in HDFS. In such a case, reducing the number of pattern matchings becomes less effective since the overhead of transferring facts over the network is much larger. Our current implementation does not focus on reducing the size of the data written to HDFS (we use no compressing functions). More study on this issue remains future work.

5 Conclusion

This paper proposed methods for implementing a rule system using the MapReduce framework. Our experimental results indicated the effectiveness of our proposed methods for speeding up in distributed environments. Reducing the number of MapReduce jobs is especially important to shorten the execution time. In addition, network I/Os between nodes can potentially become a bottleneck when the number of facts increases.

The rules and facts used in the experiments were rather simple. We only experimented with one set of rules and plan to continue our experiments with different types of rules and facts.

One of the bottlenecks in executing a rule system is calculating the join of the partial matches of a rule. To further optimize rule execution, we must analyze the interrelation between rules and define the reduce and map tasks accordingly. In addition, in the current implementation, a map task output's key is a rule-id. Thus, the number of parallel tasks in a reduce phase is limited by the number of rules. Another future issue is using a proper key to further exploit more parallelism.

References

1. Apache hadoop, `http://hadoop.apache.org/`
2. Bădică, C., Braubach, L., Paschke, A.: Rule-based distributed and agent systems. In: Bassiliades, N., Governatori, G., Paschke, A. (eds.) RuleML 2011 - Europe. LNCS, vol. 6826, pp. 3–28. Springer, Heidelberg (2011)
3. Cao, B., Yin, J., Zhang, Q., Ye, Y.: A MapReduce-based architecture for rule matching in production system. In: IEEE Second International Conference on Cloud Computing Technology and Science (CloudCom), pp. 790–795 (2010)

4. Dean, J., Ghemawat, S.: MapReduce: Simplified data processing on large clusters. In: OSDI 2004: 6th Symposium on Operating Systems Design and Implementation, pp. 137–150 (2004)
5. Forgy, C.: Rete: A fast algorithm for the many pattern/many object pattern match problem. Artificial Intelligence 19(1), 17–37 (1982)
6. Husain, M., McGlothlin, J., Masud, M., Khan, L., Thuraisingham, B.: Heuristics-based query processing for large RDF graphs using cloud computing. IEEE Transactions on Knowledge and Data Engineering 23(9), 1312–1327 (2011)
7. Ishida, T.: Parallel rule firing in production systems. IEEE Transactions on Knowledge and Data Engineering 3(1), 11–17 (1991)
8. Kifer, M.: Rule interchange format: The framework. In: Calvanese, D., Lausen, G. (eds.) RR 2008. LNCS, vol. 5341, pp. 1–11. Springer, Heidelberg (2008)
9. Myung, J., Yeon, J., Lee, S.G.: SPARQL basic graph pattern processing with iterative MapReduce. In: Proceedings of the 2010 Workshop on Massive Data Analytics on the Cloud (MDAC 2010), pp. 6:1–6:6 (2010)
10. Ravindra, P., Hong, S., Kim, H., Anyanwu, K.: Efficient processing of RDF graph pattern matching on MapReduce platforms. In: Proceedings of the Second International Workshop on Data Intensive Computing in the Clouds, DataCloud-SC 2011, pp. 13–20 (2011)
11. RDFgrid: Map/Reduce-based Linked Data Processing with Hadoop, `http://rdfgrid.rubyforge.org/`
12. Tachmazidis, I., Antoniou, G., Flouris, G., Kotoulas, S.: Scalable nonmonotonic reasoning over RDF data using MapReduce. In: Proceedings of the Joint Workshop on Scalable and High-Performance Semantic Web Systems, pp. 75–90 (2012)
13. Urbani, J., Kotoulas, S., Maassen, J., Harmelen, F.V., Bal, H.: WebPIE: A web-scale parallel inference engine using MapReduce. Web Semantics: Science, Services and Agents on the World Wide Web 10, 59–75 (2012)
14. W3C Recommendation: RIF Core Dialect, 2nd edn. (2013), `http://www.w3.org/TR/rif-core/`

Author Index

Printed in the United States
By Bookmasters